T0344682

Advances in Sintering
Science and Technology II

Advances in Sintering Science and Technology II

Ceramic Transactions, Volume 232

A Collection of Papers Presented at The International Conference on Sintering 2011, August 28—September 1, Jeju, Korea

Edited by
Suk-Joong L. Kang
Rajendra Bordia
Eugene Olevsky
Didier Bouvard

A John Wiley & Sons, Inc., Publication

Published by John Wiley & Sons, Inc., Hoboken, New Jersey.
Published simultaneously in Canada.

For general information on our other products and services or for technical support, please contact our
Customer Care Department within the United States at (800) 762-2974, outside the United States at
(317) 572-3993 or fax (317) 572-4002.

Wiley also publishes its books in a variety of electronic formats. Some content that appears in print may
not be available in electronic formats. For more information about Wiley products, visit our web site at
www.wiley.com.

Library of Congress Cataloging-in-Publication Data is available.

ISBN: 978-1-118-27375-3
ISSN: 1042-1122

Printed in the United States of America.

10 9 8 7 6 5 4 3 2 1

Contents

MICROSTRUCTURAL EVOLUTION AND PHYSICAL PROPERTIES

UNCONVENTIONAL SINTERING PROCESSES

Preface

This issue of the Ceramic Transactions includes a number of papers presented at the International Conference on Sintering 2011, which was held on Jeju Island, Republic of Korea, on August 26—September 1, 2011. The meeting was chaired by Professors Suk-Joong L. Kang, Rajendra Bordia, Eugene Olevsky, and Didier Bouvard.

This was the sixth meeting in a series that started in 1995 as a continuation of the well-known cycle of conferences on sintering and related phenomena organized by G. Kuczynski which ran from 1967 to 1983. The first five meetings in this reestablished series of conferences were held at Pennsylvania State University in 1995, 1999 and 2003; in Grenoble, France in 2005; and in San Diego, California in 2008.

Sintering 2011 brought together more than 260 participants from 27 countries, fostering a high level of scientific interaction and creating an atmosphere of broad international collaboration. The meeting included participants from North and Central America, Europe (both Eastern and Western), Asia, Australia and Africa.

The conference demonstrated the advances that have been made in the areas of the multi-scale modeling of densification and in microstructure development and promoted a better understanding of the processing of complex systems (multi-layered, composite and reactive systems). Concerning sintering technology, innovative approaches such as field-assisted sintering attracted the attention of the materials processing community. Other timely topics included the sintering and microstructural development of nanostructured materials, and the sintering of bio- and energy applications-related materials.

Twenty five papers of Sintering 2011 authors were published in special issues of the Journal of the American Ceramic Society and of the Journal of Materials Science (July 2011). The American Ceramic Society is publishing approximately 20 papers presented at this meeting in this volume of the Ceramic Transactions. Together, these forty five papers cover the rich diversity of the sintering science and technology topics, which were presented at the conference. Both the conference participants and organizers had to meet numerous deadlines to enable the timely publication of this volume and of the special issues of the Journal of the American Ceramic Society and of the Journal of Materials Science.

We hope that this collection of articles will be an important contribution to the literature, and we are looking forward to seeing you at future Sintering conferences.

SUK-JOONG L. KANG, RAJENDRA BORDIA, EUGENE OLEVSKY, AND DIDIER BOUVARD
Sintering 2011 Co-Chairs

Powder Synthesis
and Sintering

DEPOSITION OF PLATINUM NANOPARTICLES ONTO COPPER FOILS BY ELECTROPHORESIS: A STUDY OF THE SINTERING DYNAMICS AT THE PLATINUM-COPPER INTERFACE

Deborah C. Blaine[*], Alexander Ilchev[†], Leslie Petrik[†], Patrick Ndungu[‡], and Alexander Nechaev[†]

[*]Mechanical and Mechatronics Engineering, Stellenbosch University, South Africa
[†]Environmental Nanosciences, University of the Western Cape, South Africa
[‡]Department of Chemistry, University of KwaZulu-Natal, South Africa

ABSTRACT

The effect of decreasing the size of platinum to the nanoscale for the copper-platinum alloy system was investigated while coupled to the research on the electrophoresis of platinum nanoparticles. It was observed that the temperature for the nanoPt-Cu eutectoid transformation decreased to 300°C compared to the value of 418°C for a bulk Pt-Cu system. HRTEM and XRD results showed that sintering between the Pt nanoparticles and copper surface begins at temperatures as low as 200°C. Surface diffusion was dominant at 200°C while bulk diffusion became dominant at 300°C, at which stage the sintered product had formed a tri-phasic system of pure copper, Pt-Cu alloy and Pt nanoparticles which were sintered to the alloy. No platinum nanoparticles remained after sintering at 400°C, having all diffused into the Pt-Cu alloy. Sintering above 300°C saw the appearance of the Cu_3Pt phase. HRTEM imaging of the particles found that there was an increase in particle size from 2.5 nm at 25°C to 0.5 microns at 400°C. These results were compared to particle size calculations using XRD measurements. The particle geometry changed from spherical at 25°C to cubic at and above 300°C.

INTRODUCTION

Nanotechnology is defined as any technology performed on a nanoscale that has applications in the real world.[1] It encompasses any device, natural or synthetic, that has at least one of its dimensions on the length scale of 0.1 – 100 nm. Such devices include thin films, nanolayered materials and membranes (one dimension on the nanoscale), nanotubes, nanorods, nanowires and nanofibres (two dimensions on the nanoscale), as well as nanoparticles, nanospheres, quantum dots, micelles and dendrimers (all three dimensions on the nanoscale.) These are the basic building blocks in nanotechnology and the integration of such components is the basis for the design of nanostructured systems. Perhaps the most exciting discovery in this field was the quantum size effect[1] where as a material's dimensions are reduced to the nanoscale, its properties begin to gradually shift from those of the bulk to novel ones which are better explained by quantum mechanics. Understanding the parameters which control the synthesis outcome and inherent properties of these components allows for the tailoring of novel technological devices which allow the fantastic idea of being able to control chemical systems, on a molecular or even atomic level, to come to life.

The use and development of platinum based materials is becoming ever more crucial in modern technology each day. The application of platinum and its alloys in the field of heterogeneous catalysis is well known and, in many cases, held as a standard. Fuel cell research, as well as other aspects of the hydrogen economy, is mostly based on developing such materials.[2] Additionally, catalytic converters, used to reduce emissions in automobiles and for catalytic cracking in the petroleum industry, have been dependant on platinum alloys since their commercialisation.[3] Other application for platinum in the catalysis industry include the production of chemicals, such as nitric acid[4], hydrogen cyanide[5] and various others hydrocarbons relevant to the petrochemical industry.[6,7]

The selectivity and activity of each process can be modified by using alloys of the different precious metals with each other and/or with other transition metals.[8, 9] Work by both Chandler et al.[10] as well as Weihua et al.[11] have shown that Pt-Cu bimetallic nanoparticles also show superior catalytic

3

properties compared to pure platinum. Koh *et al.*[12] and Mani *et al.*[13] both reported on the development of dealloyed Pt-Cu nanoparticles with a near-surface alloy structure as well as a high degree of porosity which not only improved catalyst performance by up to 4 times that of commercial Pt-C but also improved catalyst tolerance to poisoning. Pure platinum has shown poor performance as a catalyst for the oxygen reduction reaction (ORR). The reduction of oxygen in aqueous media by platinum catalysts has been extensively studied for application in fuel cells by Antolini *et al.*[9,14], Bell[15], Luo *et al.*[16,17], Regalbuto[18], Koh *et al.*[12], Mani *et al.*[13] as well as Wang *et al.*[19]. They have shown that binary alloys of platinum with other metals such as cobalt, chromium, vanadium, titanium and copper has shown a marked increase in catalytic activity towards the oxygen reduction reaction.

Van der Biest *et al.*[20] as well as Bersa *et al.*[21] stated that EPD has been successfully employed in the formation of wear-resistant and anti-oxidant coatings, functional films in microelectronic devices and solid oxide fuel cells, membranes, sensors as well as composite and bioactive coatings for medical implants. Teranishi *et al.*[22] prepared 2D nanoarrays of platinum nanoparticles on carbon coated copper grids using EPD. Very little work exists for the EPD of platinum[22-25] hence this research offers insight into possible novel applications for the EPD process of platinum.

By developing a better understanding of the sintering dynamics of the Pt-Cu biphasic system, novel processes for the self-assembly of superior materials of the types mentioned above could be designed. The goal in this research was to determine the effects that reducing the platinum phase to the nanoscale had on the properties of the system.

Abe *et al.*[26] reported that for this system, a eutectoid transformation occurs at 418°C. In theory, reducing either of the phases to the nanoscale would decrease this limit for eutectoid formation. The parameters controlling the formation of CuPt and Cu_3Pt alloys as well as the diffusion characteristics between the Pt and Cu phases was considered by developing a two phase system with a defined boundary between each phase. This was achieved by depositing the platinum nanophase onto a copper foil using electrophoretic deposition (EPD).

EXPERIMENTAL PROCEDURES
Pt Nanoparticle Synthesis and Dispersion

A dispersion of platinum nanoparticles was prepared by the ethylene glycol method[27]. To prepare 200 ml of Pt nanoparticles dispersed in a 1:1 (v/v) mixture of anhydrous acetone and ethanol (AcEt), the following procedure was applied: 1g of NaOH (25 mmol) was dissolved in 50 ml of ethylene glycol (EtGly) at 100°C under N_2. 1g of H_2PtCl_6 (2.44 mmol) was also dissolved in 50 ml of EtGly at room temperature under nitrogen. The two solutions were then mixed together, heated up to 160°C under nitrogen and allowed to react with constant stirring for 2 hours. Once the reaction was complete, 100 ml of ultrapure water was added followed by 25 ml of 1M HCl. The mixture was centrifuged for 30 minutes at 5000 rpm after which the supernatant liquid was discarded. Thereafter, 200 ml of ultrapure water as well as 10 ml of 1M HCl was added to the precipitate, the mixture was shaken well to redisperse the platinum particles. The dispersion was centrifuged again for the same time at the same speed. After centrifugation, the supernatant liquid was discarded again and the platinum precipitate was redispersed in 200 ml of AcEt.

Copper Electrode Preparation

Copper electrodes for the EPD process, with dimensions of 50 mm × 10 mm, were cut from a 5.5 g copper foil of 0.25 mm thickness (supplier: Sigma Aldrich). The copper foil was sanded, sequentially, with 400 grit followed by 600 grit SiC paper, using an active oxide suspension (Struers OP-S suspension), and then polished using cotton wool only, soaked in the suspension. The electrodes were then rinsed with ultrapure water, followed by acetone and finally they were immersed in acetone and sonicated for 15 minutes. The electrodes were then dried with cotton wool.

Electrophoretic Deposition of the Pt Nanoparticles

The EPD apparatus consisted of a DC power supply (supplier: TDK-Lambda, model: Genesys™ GEN600:1.3, 0-600 Volts DC and 0-1.3 Amps DC range, 72 mV and 0.26 mA accuracy) which was connected to a cylindrical brass anode holder and a hollow, cylindrical, stainless steel cathode. The electrodes were housed in a Pyrex container so that the anode rested inside the cathode. Figure 1 shows a schematic diagram of EPD cell.

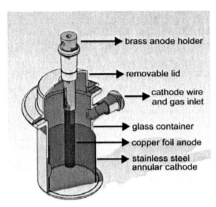

→ brass anode holder

→ removable lid

→ cathode wire and gas inlet

→ glass container

→ copper foil anode

→ stainless steel annular cathode

Figure 1. Schematic of the EPD cell.

A set of trials was performed to determine the deposition flux as a function of voltage and deposition time. Each trial was performed in air and 25 ml of the platinum dispersion at a concentration of 1.8 g/l was used. 12 trials with a potential range of 2 – 10 volts and a time range of 5 – 30 minutes were performed to gain an understanding of the deposition kinetics for the system. Table 1 gives the EPD parameters tested.

Table I Pt nanoparticle deposition flux at different EPD trial parameters: time at voltage

Deposition time (mins)	Deposition flux (mg/mm^2)		
5	0.0118	0.0210	0.0254
10	0.0150	0.0235	0.0314
20	0.0192	0.0257	0.0327
30	0.0200	0.0259	0.0331
Deposition voltage (V)	2	5	10

To prepare a sample deposit for sintering, 100 ml of the platinum dispersion was diluted with an equal volume of acetone. This dispersion, at a concentration of 0.9 g/l, was then used in the EPD cell to deposit the platinum nanoparticles as a film on the copper foil electrode. The cell was set to an applied voltage of 5 V and deposition occurred for 10 minutes.

Sintering the Pt Deposits

Sintering of the deposits was achieved in a vertical tube furnace. The copper electrode, onto which the Pt nanoparticles had been deposited, was suspended at the mouth of the furnace with a tungsten wire which ran through the furnace. The furnace was heated to the required temperature while being purged by argon gas. The electrode was then hoisted into the furnace with the tungsten wire and kept under flowing argon while the sintering took place. After 15 minutes, the furnace was rapidly purged with argon while the electrode was lowered out of the furnace. Deposit samples were sintered at 100°C, 200°C, 300°C and 400°C.

RESULTS

Synthesis of the Dispersed Pt Nanoparticles

The dispersion prepared was analyzed by HRTEM (high resolution transmission electron microscopy) using a Tecnai F20 field emission transmission electron microscope with coupled EDX (energy dispersive x-ray spectroscopy) capabilities, and XRD (X-ray diffraction) analysis using a PANalytical X'Pert Pro multipurpose diffractomator. The JCPDS (Joint Committee on Powder Diffraction Standards) database was used to characterize the XRD data collected.

The concentration of Pt in dispersion was measured as follows. Three 10 ml samples of the dispersion were each placed in pre-weighed glass vials. The mass of the dispersion in each sample was calculated by weighing the vials after the samples were added. Thereafter, the solvent was evaporated and the vials holding the precipitate only were weighed. The mass of the precipitate multiplied by 100 gives the concentration (in grams per litre) of Pt in the dispersion. The average concentration was taken as the concentration of Pt in the dispersion.

Figure 2(a) shows the HRTEM image of the Pt nanoparticles synthesized by the above procedure and well dispersed in the AcEt solvent mixture. The image shows that the particles are crystalline with an approximate mean particle size of 2 nm. A histogram of the particle size distribution of the dispersed Pt nanoparticles is shown in Figure 2(b). The data was collected by measuring and recording the particles sizes of 51 particles from eight SEM images by hand.

Figure 2(c) shows the XRD pattern for the dispersed Pt nanoparticles. The 2θ peaks of 39.1°, 45.6°, 66.6° and 81.5° in Figure 2(c) are indicative of face centred cubic (fcc) platinum, with the lattice planes for the peaks corresponding to the (111), (200), (220) and (311) planes, respectively. The mean particle diameter, calculated from the XRD spectrum, Figure 2(c), using the Sherrer equation, is 2.2 nm which was in close agreement with the HRTEM results.

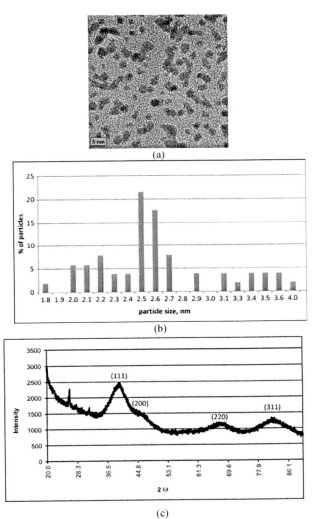

Figure 2. Imaging and properties of the Pt nanoparticles synthesised: (a) HRTEM image, (b) particle size distribution measured from HRTEM image, and (c) XRD pattern.

EPD of the Pt Nanoparticles onto the Copper Foil Electrodes

The data gathered in the EPD trials was plotted in Figure 3 as deposition flux with respect to deposition time for each applied voltage.

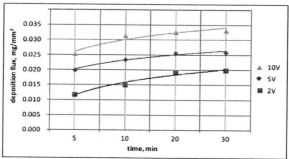

Figure 3. Deposition flux vs. deposition time for EPD trials.

Looking at the graph in Figure 3, the slope of each curve at each applied voltage appears to be similar for all the applied voltages, hence the polarization of the electrode in this system is assumed to be linear for the potential window of 2 – 10 V.

By inspecting the deposits with the naked eye, only the deposits prepared with an applied voltage of 5 V or higher appeared to cover the entire surface of the electrode in contact with the dispersion. Using SEM analysis, performed with a ZEISS EVO MA15VP scanning electron microscope, the thinnest and most uniform film was obtained by running the EPD process at an applied potential of 5 V for 5 minutes. Figure 4(a) shows the SEM image of the deposit prepared on a copper electrode by EPD for 5 minutes at an applied voltage of 5 V. The effect of lowering the dispersion concentration was studied by diluting the dispersion. This was achieved by adding 100 ml of acetone to 100 ml of the prepared dispersion, effectively reducing the Pt concentration in the dispersion to 0.9 g/l. By reducing the amount of ethanol in the solvent mixture as well as the concentration of platinum nanoparticles in dispersion, the deposition flux was further lowered without affecting the uniformity of the deposit significantly, as shown in Figure 4(b). The deposition time was doubled for the diluted dispersion to ensure a continuous platinum film was formed on the copper surface. As the deposition with the diluted dispersion for 10 minutes at 5V gave the best result, these parameters were used to prepare electrodes for the sintering analysis.

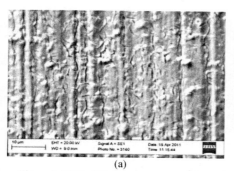

(a)   (b)

Figure 4. SEM of Pt deposit on copper electrode after EPD for (a) 5 minutes at 5 V with undiluted dispersion, and for (b) 10 minutes at 5 V with the diluted dispersion.

Sintering of the Pt Deposits

The sintered deposits were analysed by SEM, EDX, XRD, and HRTEM. The microstructure in each of the deposits was investigated using a ZEISS EVO MA15VP scanning electron microscope. Figure 5 shows representative images for each sintered sample.

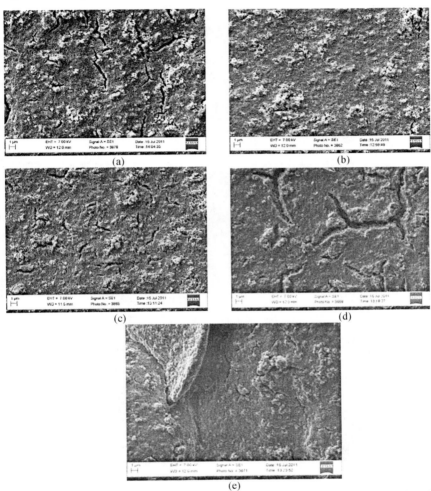

(a)

(b)

(c)

(d)

(e)

Figure 5. SEM images of Pt deposited on copper electrodes (a) at 25°C, unsintered, and after sintering at (b) 100°C, (c) 200°C, (d) 300°C, and (e) 400°C, in argon for 15 minutes each.

The cracks in the deposit before sintering, as seen in Figure 5(a), disappeared after sintering at 100°C, Figure 5(b), suggesting that the deposited particles were mobile on the copper surface. Both the HRTEM and XRD data indicate no significant change in the mean particle size of the platinum nanoparticles. Table 2 shows the XRD and HRTEM measurements of the mean particle sizes in the deposits.

Table II. XRD and HRTEM measurements of the mean particle size in the deposits

Sintering Temperature (°C)	Mean particle size: HRTEM (nm)	Particle size variation: HRTEM (nm)	Mean particle size: XRD Scherrer (nm)
25	2.5	1.8 - 4.0	2.56
100	2.8	2.0 - 4.0	2.52
200	3.9	2.5 - 5.0	3.96
300	43	2.5 - 150.0	11.72
400	100	10.0 - 600.0	13.83

The XRD spectrum, Figure 6, for the deposit sintered at 100°C also shows that no shift in the d-spacing for the platinum crystal planes has taken place since there is no shift in the 2θ values for the Pt peaks. Hence no alloying between the platinum deposit and the copper electrode had taken place after heating at 100°C.

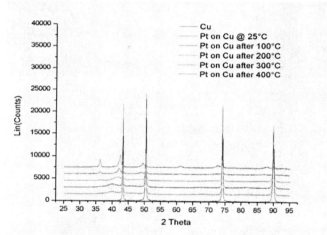

Figure 6. XRD spectra for the copper electrodes as well as the Pt deposits prepared by EPD on the copper electrodes.

HRTEM imaging of the platinum deposit after sintering at 100°C, Figure 7(b), indicates that no necking between the platinum nanoparticles had occurred at 100°C. Therefore, it can be concluded that no significant surface diffusion was taking place between the particles. The disappearance of the cracks in the deposit can be attributed to settling of the Pt nanoparticles as the residual solvent and water evaporated from the deposit.

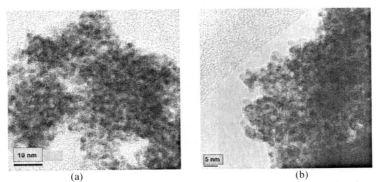

(a)   (b)

Figure 7. HRTEM image of Pt deposit scraped off the copper electrode (a) before sintering, and (b) after sintering at 100°C.

Fine cracks reappeared on the deposit surface after sintering at 200°C, shown in Figure 5(c). This was indicated the formation of a densification gradient between the deposit and the copper surface, as well as immobilization of the deposit due to bonding with the copper substrate. In other words, sintering of the platinum nanoparticles must have begun at this temperature. Furthermore, the shift in the 2θ value for the [111] peak in the XRD spectrum of the platinum deposit from 40.0° to 41.6° is indicative of a decrease in the d-spacing of the [111] planes for the platinum phase indicating the onset of alloying. The other peaks had signals which were too weak to be detected. The mean particle size from both the XRD and HRTEM data, Figure 6 and Figure 8(a), was calculated to be 3.9 nm in both cases, indicating that very little coarsening had occurred. Densification of the deposit was occurring dominantly via surface diffusion at this temperature.

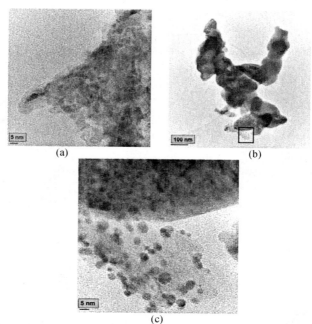

Figure 8. HRTEM image of Pt deposit after sintering at (a) 200°C, and (b) 300°C, with magnified imaging of the area in the block shown in (c).

After sintering at 300°C, both HRTEM imaging, Figure 8 (b) and (c), and XRD analysis, Figure 6, unambiguously indicate both sintering as well as alloying between the Pt nanoparticles and the Cu surface. The HRTEM images show significant particle growth as well as distinct necking between the particles. Upon closer inspection, Figure 8(c), the surface of most of the sintered particles is dispersed with pure platinum nanoparticles, which are sintered to the main alloy. The XRD spectrum for this deposit sample showed a further increase in the 2θ value for the [111] peak from 41.62° to 42.22°, closer to the peak characteristic of the Cu_3Pt phase. The peak intensities also rose, indicating an increase in the amount of alloy forming. The [200] peak was now also visible. The peaks at 2θ = 36.2° and 61.0° are characteristic for the [111] and [220] peaks, respectively, for Cu_2O. This suggested that either some oxygen was being introduced into the furnace during sintering or that some organic stabilizing agents on the Pt nanoparticles in the dispersion, such as glycolate ions, were being retained in the deposit and were reacting with the deposit at 300°C to release oxygen. Due to the low concentration of Cu_2O that had formed, no HRTEM images of its presence in the deposits could be taken.

At this point, a triphasic system had formed including the pure copper substrate, sintered Pt-Cu alloy and the unsintered Pt nanoparticles. Here, the correlation between the mean particle size calculated using the HRTEM and XRD data was lost. This was due to the peak overlap between the Pt and Cu_3Pt peaks in the XRD spectrum, i.e. the Cu_3Pt peaks were interfering with the much less intense

Pt peaks, hence the mean particle size of the Pt nanoparticles on the Pt-Cu alloy could not be calculated from the XRD data for that sample. The polycrystalline nature of the eutectoid formation was another factor which led to the Scherrer equation for the XRD data for the sample sintered at 300°C effectively calculating the mean grain size of the Pt-Cu eutectoid phase. Looking at the trend for the variation in particle size for each deposit, shown in Table 2, it is confirmed that sintering at temperatures of 300°C and above causes a loss in the correlation between the calculated values for the mean particle sizes of the samples when using HRTEM and XRD data. The HRTEM data is considered less ambiguous for such multiphasic systems, though it is more tedious to analyse. The XRD data is still useful in predicting the size of the grains in a polycrystalline solid. The small size of the grains for the samples sintered at 300°C and 400°C is characteristic of a rapidly cooled eutectoid.

Figure 9 shows an HRTEM image of the Pt deposit after sintering at 400°C. Compared to Figure 8(a) and 8(b), it shows that the deposit has sintered to the point where the previous particle boundaries are no longer apparent and the deposit now appears as a continuous film. There is no indication of discrete Pt nanoparticles on the deposit surface, as was the case in Figure 8(c). Table III gives the EDX analysis on the deposit surface measured at several points, reported as the average of the measurements. There is no pure Pt present at any of the points analysed. It is surmised that the Pt has diffused below the surface, alloying with the Cu substrate at subsurface level.

XRD analysis, Figure 6, indicated that the most prominent alloy within the nanoparticle structures was Cu_3Pt. It also indicated that the amount of Cu_2O in the sample increased.

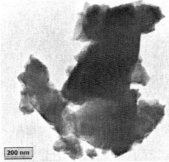

200 nm

Figure 9. HRTEM images of Pt deposit scraped off the copper electrode after sintering at 400°C.

Table III. EDX data for the deposit sintered at 400°C

Element	Weight %	Atomic %	Uncert. %	Detector correction	k-factor
C(K)	9.10	26.25	0.12	0.26	3.94
O(K)	14.74	31.90	0.11	0.49	1.974
Ni(K)	7.81	4.61	0.07	0.99	1.511
Cu(K)	68.33	37.23	0.20	0.99	1.667
Pt(L)	0.00	0.00	100	0.75	5.547

A change in the particle geometry from spherical at 25°C to mostly cubic was noted to begin at 300°C and it became more pronounced at 400°C, as can be seen in Figure 9. This suggested that the Cu_3Pt system favours growth of the [100] crystal face. This deduction was in contradiction to the XRD results which showed a higher diffraction intensity for the lower surface energy inducing [111] crystal planes. This suggests that a different crystal structure exists at the deposit surface compared to the subsurface level. If a temperature gradient had forms from the deposit surface down, then crystal growth would be kinetically favoured at the surface, where the phase transformation was more rapid, resulting in the formation of the less kinetically hindered [100] crystal faces. These faces also contain fewer platinum atoms than the [100] crystal planes, meaning that they would be more thermodynamically feasible to form at the surface during the diffusion of the Pt into the Cu substrate. Similar behaviour has been reported for Pt evaporated onto a Cu single crystal [111] face and heated above 500 K (227°C).[28] For contrasting reasons, the subsurface level would have grains more randomly oriented, with the most thermodynamically stable [111] planes being the predominant grain boundaries. The high magnification HRTEM image of the sample sintered at 400°C, Figure 10, shows the lattice fringes within the sample indicating the formation of a polycrystalline phase.

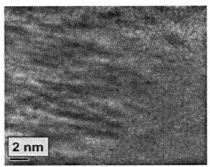

Figure 10. HRTEM image of deposit sintered at 400°C.

CONCLUSIONS AND FUTURE WORK

This research demonstrated that Pt nanoparticles with a narrow particle size distribution are formed easily and remain in dispersion indefinitely when prepared by the ethylene glycol synthesis method. The dispersions prepared were also stable throughout the timeline of the research. It was also demonstrated that the EPD process can be applied to the dispersions of Pt nanoparticles even in a low electrolyte media, though a small amount of reactive solvent or redox initiator aids in the EPD kinetics, as does the polarizability of the electrode surface. It was confirmed that the dynamics of the EPD process were retained in the setup used and deposits with various densities, thicknesses and coherencies could be prepared by only altering the EPD parameters, applied potential, deposition time and dispersion concentration. By developing the EPD process a better control of the deposit density, thickness and coherency can be obtained. Factors such as the solvent composition, use of stabilizing agents and particle concentrations should all be looked into for improving the versatility of the process. Sintering of the platinum deposit was achieved at low temperatures compared to typical bulk platinum sintering temperatures. The eutectoid transformation, typically occurring at 418°C for bulk Pt-Cu alloys, occurred at a reduced temperature, around 300°C, for Pt nanoparticles sintered to a Cu substrate. The sintering kinetics were slow at 200°C, due to the copper existing as a bulk phase, hence sintering occurred dominantly via surface diffusion, and little Cu-Pt alloy formation occurred. The triphasic deposit, formed after sintering at 300°C, could be considered for a heterogeneous catalyst system due to the fine dispersion of platinum on the eutectic surface. Work on the sintering dynamics, through optimization of the sintering times as well as heating rates and cooling rates, could aid in improving this procedure to compete with current state of the art synthesis methods for catalysts and hard disk platters. Determining whether the oxygen responsible for the Cu_2O formation came from the atmosphere or from occluded organic impurities in the deposit needs to be determined and circumvented in order to ensure purity of the structures formed by this procedure. While most of the work in this article was conceptual, testing the sensitivity of this system to scaling up should be considered as the procedures used in this research are relatively inexpensive.

ACKNOWLEDGEMENTS

This work was presented at The International Conference on Sintering 2011, 28 August – 2 September 2011, Jeju, Korea. It was funded by the National Research Foundation (South Africa) under their Nanotechnology Flagship Programme, Project "Nano-architecture in the beneficiation of platinum group metals."

REFERENCES
[1] B. Bhushan, Springer Handbook of Nanotechnology, Springer, USA (2004).
[2] J. Rifkin, The Hydrogen Economy: The Creation of the Worldwide Energy Web and the Redistribution of Power on Earth. J.P. Tarcher/Putnam, USA (2002).
[3] S.T. Gulati, New Developments in Catalytic Converter Durability, *Studies in Surface Science and Catalysis* 71, 481-507 (1991).
[4] W. Ostwald, Improvements in the Manufacture of Nitric Acid and Nitrogen Oxides, *Patent GB 190200698(A)* (1902).
[5] R.J.H. Voorhoeve, C.K.N. Patel, L.E. Trimble, and R.J. Kerl, Hydrogen Cyanide Production During Reduction of Nitric Oxide over Platinum Catalysts: Transient Effects, *Science*, 200 , 761-763 (1978).
[6] J.G. Speight, Synthetic fuels handbook: properties, process, and performance, McGraw-Hill Professional, USA (2008).
[7] N.H. Sager, and R.M.L. Pouteau, The Production of Heavy Water: Hydrogen-Water Deuterium Exchange Over Platinum Metals on Carbon Supports, *Platinum Metals Rev.*, 19 (1), 16-21 (1975).
[8] S. Nishimura, Handbook of heterogeneous catalytic hydrogenation for organic synthesis, J. Wiley, USA (2001).
[9] E. Antolini, Platinum-based ternary catalysts for low temperature fuel cells Part I. Preparation methods and structural characteristics, *Applied Catalysis B: Environmental*, 74, 324-336 (2007).
[10] B.D. Chandler, A.B. Schabel, L.H. Pignolet, Preparation and Characterization of Supported Bimetallic Pt-Au and Pt-Cu Catalysts from Bimetallic Molecular Precursors., *J. Catal.*, 193 (2), 186-198 (2000).
[11] W. Weihua, T. Xuelin, C. Kai, and C. Gengyu, Synthesis and characterization of Pt–Cu bimetallic alloy nanoparticles by reverse micelles method, *Colloids and Surfaces A: Physicochem. Eng. Aspects*, 273, 35-42 (2006).
[12] S. Koh, and P. Strasser, Electrocatalysis on bimetallic surfaces: modifying catalytic reactivity for oxygen reduction by voltammetric surface dealloying, *J Am Chem Soc.*, 129, 12624-12625 (2007).
[13] P. Mani, R. Srivastava, and P. Strasser, Dealloyed Pt–Cu Core–Shell Nanoparticle Electrocatalysts for Use in PEM Fuel Cell Cathodes, *J. Phys. Chem. C*, 112, 2770-2778 (2008).
[14] E. Antolini, R.R. Passos, and E.A. Ticianelli, Electrocatalysis of oxygen reduction on a carbon supported platinum–vanadium alloy in polymer electrolyte fuel cells, *Electrochimica Acta*, 48, 263-270 (2002).
[15] A.T. Bell, Catalysis looks to the future. National Academies Press, USA, (1992).
[16] J. Luo, L. Han, N. Kariuki, L. Wang, D. Mott, C.J. Zhong, and T. He, Synthesis and Characterization of Monolayer-Capped PtVFe Nanoparticles with Controllable Sizes and Composition, *Chem. Mater.*, 17, 5282-5290 (2005).

[17] J. Luo, L. Wang, D. Mott, P.N. Njoki, N. Kariuki, C.J. Zhong, and T. He, Ternary alloy nanoparticles with controllable sizes and composition and electrocatalytic activity, *J. Mater. Chem.,* **16**, 1665-1673 (2006).

[18] J.R. Regalbuto, Catalyst preparation: Science and Engineering, CRC Press, USA (2006).

[19] Y. Wang, J. Ren, K. Deng, L. Gui, and Y. Tang, Preparation of Tractable Platinum, Rhodium, and Ruthenium Nanoclusters with Small Particle Size in Organic Media, *Chem. Mater.,* **12**, 1622-1627 (2000).

[20] O.O. Van der Biest, and L.J. Vandeperre, Electrophoretic Deposition of Materials, *Annu. Rev. Mater. Sci.,* **29**, 327-352 (1999).

[21] L. Besra, and M. Liu, A review on fundamentals and applications of electrophoretic deposition (EPD), *Progress in Materials Science,* **52**, 1-61 (2007).

[22] T. Teranishi, M. Miyake, and M. Hosoe, Formation of Monodispersed Ultrafine Platinum Particles and their Electrophoretic Deposition on Electrodes, *Adv. Mat.,* **9** (1), 65-67 (1997).

[23] T. Teranishi, M. Hosoe, T. Tanaka, and M. Miyake, Size Control of Monodispersed Pt Nanoparticles and Their 2D Organization by Electrophoretic Deposition, *J. Phys. Chem. B,* **103**, 3818-3827 (1999).

[24] X. Yin, Z. Xue, and B. Liu, Electrophoretic deposition of Pt nanoparticles on plastic substrates as counter electrode for flexible dye-sensitized solar cells, *J. Power Sources* **196** (4), 2422-2426 (2011).

[25] J.S. Zheng, M.X. Wang, X.S. Zhang, Y.X. Wu, P. Li, X.G. Zhou, and W.K. Yuan, Platinum/Carbon Nanofiber Nanocomposite Synthesized by Electrophoretic Deposition as Electrocatalyst for Oxygen Reduction, *J. Power Sources,* **175** (1), 211-216 (2008).

[26] T. Abe, B. Sundman, and H. Onodera, Thermodynamic Assessment of the Cu-Pt System, *J. Phase Equilibria and Diff.,* **27**(1), 5-13 (2006).

[27] A. Ilchev, Platinum group alloy nanoparticle architecture and their electrophoretic deposition, MSc thesis, University of the Western Cape, South Africa (2011).

[28] U. Schröder, R. Linke, J-H. Boo, and K. Wander, Adsorption properties and formation of Pt/Cu surface alloys, *Surface Sci.,* **352-354**, 211-217 (1996).

PRESSURELESS SINTERING AND PIEZOELECTRIC PROPERTIES OF MECHANOCHEMICALLY SYNTHESIZED $K_{0.5}Na_{0.5}NbO_3$ POWDER COMPACTS

Jung-Yeul Yun[1,*], Si-Young Choi[1], Min-Soo Kim[2], and Suk-Joong L. Kang[3]

[1]Fuctional Materials Division, Korea Institute of Materials Science, Changwon, 641-010, Republic of Korea
[2]Advanced Materials & Application Research Laboratory, Korea Electrotechnology research Insitute, Changwon, 641-120, Republic of Korea
[3]Department of Materials Science and Engineering, Korea Advanced Institute of Science and Technology, Daejon, 305-701, Republic of Korea

*Corresponding author, E-mail: yjy1706@kims.re.kr

ABSTRACT

Alkali sodium niobate, $K_{0.5}Na_{0.5}NbO_3$ (KNN), is one of the promising substitutions for lead-based piezoelectric materials . Fabrication of dense KNN by pressureless sintering, however, is difficult because of its low phase stability at high temperatures and the volatility of the alkali elements. We attempted pressureless sintering of KNN ceramics prepared from mechanochemically synthesized powder with the crystallite size of 10-20 nm. The powder compacts was fully densified after 12 h sintering at 1100°C in oxygen. The sintered KNN exhibited a piezoelectric constant of 119 pC/N and a coupling factor of 0.41. These values are superior to those reported in previous investigations.

I. INTRODUCTION

Lead-based piezoelectric ceramics have widely utilized in mechanical sensors, actuators and other electronic applications because of their excellent piezoelectric properties. The harmfulness of lead to the environment, however, has led a strong drive to develop lead-free compounds to replace lead-based piezoelectric ceramics.[1-4] Among several candidates of lead-free materials, alkali niobate-based materials, typically KNN, have been considered promising alternatives to lead-based piezoelectric materials.[5-7] Fabrication of dense KNN ceramics, however, has been difficult because of its low sinterbility in conventional processing. In pressureless sintering, raising the sintering temperature has commonly been considered to be a solution to prepare dense KNN ceramics. The phase stability of KNN, however, is limited to 1140 °C. In addition, the volatility of potassium and its high reactivity with moisture impede the densification of KNN in conventional sintering.[1,8] In order to overcome the low sinterability of KNN, pressure-assisted sintering[9-10] has been attempted. This technique, however, increases the production cost. Probably the simplest solution is the reduction of powder size, which can increase the sinterability.[11] KNN powders are commonly produced by the conventional solid-state reaction process and their average particle size is quite large, in the order of 0.1 μm. To produce fine KNN powders, chemistry-based methods, such as hydrothermal[12] and microemulsion processes,[13-15] have also been attempted. Those methods, however, require complex treatments of high purity and expensive starting materials.[12-15]

Our previous investigation[16] demonstrated the preparation of fine $NaNbO_3$ powder by the mechanochemical method. As a continuation, this investigation attempts mechanochemical synthesis of sinterable fine KNN powder using a powder mixture of carbonates and oxides. The synthesized powder was easily consolidated by pressureless sintering. The sintered samples also exhibited excellent piezoelectric properties.

II. EXPERIMENTAL

Two methods, conventional solid-state reaction (SS) method and mechanochemical (MC) method, were used to synthesize KNN powders from commercially available K_2CO_3, Na_2CO_3 and Nb_2O_5 (Sigma-Aldrich Inc., Milwaukee, USA) powders. K_2CO_3 and Na_2CO_3 powders were dried at 250 °C for 2 h prior to mixing with other powders in order to eliminate the adsorbed water.

For the synthesis of KNN powder by the SS(Solid State Reaction) method a proportioned powder mixture was ball-milled in ethanol for 24 h using a polypropylene jar and ZrO_2 milling balls. After ball milling, the ethanol was evaporated using a hot plate and magnetic stirrer. The dried slurry was crushed in an agate mortar using a pestle and passed through a 150 μm sieve. The powder was loaded in an alumina crucible with lid and calcined at 850 °C for 5 h in air.

For the synthesis by the mechanochemical method, two different routes (route M and MA) were adopted. In route M (mechnochemical synthesis without annealing), a proportioned powder mixture was loaded in a cylindrical WC vial with WC balls and mechanochemically milled using a high energy shaker mill (SPEX mill) operated at ~1200 rpm. For the milling, ball-to-powder ratio (BPR) was 30:1 and the milling time was 7 h. In the case of route MA (mechanochemical synthesis with annealing), the mechanochemically synthesized KNN powder was annealed at 800 °C for 5 h in order to remove the unreacted carbonates. The annealed powder was wet milled for 30 h in ethanol using a planetary mill with ZrO_2 milling balls and a ZrO_2 jar. After drying on a hot plate, the powder was ground in an agate mortar and passed through a 150 μm sieve. The particle size distribution of the powders was measured using a laser particle size analyzer (HELOS particle size analysis, Sympatec GMbH, Germany).

The prepared powder was lightly pressed in a steel die of 10 mm in diameter and cold isostatically pressed at 200 MPa. The powder compacts were sintered at 1100 °C for 12 h in an oxygen atmosphere. The densities of sintered samples were measured by the Archimedes' method. The morphology of powders and the microstructure of the sintered samples were observed under a field-emission scanning electron microscope (FE-SEM) (JSM6700F, JEOL, Japan) and a transmission electron microscope (TEM) (JEM-2100F, JEOL, Japan).

To measure the piezoelectric properties, both sides of the disk samples were polished and painted with a silver paste. The samples were poled in a silicon oil bath at 150 °C for 30 min under a 30 kV/cm. The piezoelectric constant was measured with a quasi-static meter of Berlincourt type (Piezo d-meter, CADT, USA). The electromechanical coupling coefficients were determined by the resonance-antiresonance method using an impedance analyzer (HP4294A, Agilent Technologies, USA).

III. RESULTS AND DISCUSSION

Figure 1 plots the XRD patterns of KNN powders prepared by SS method (a), MC method without annealing (route M) (b), and MC method with annealing (route MA) (c). The XRD pattern in Fig. 1(a) indicates that in the detection limit of X-ray diffraction, a single phase KNN was formed by solid state reaction, as in previous studies.[17,18] Figures 1(b) and 1(c) also show that a single phase KNN was formed after mechanochemical treatment. The XRD peaks are, however, broad for the synthesized powder without annealing, suggesting that the crystals are heavily deformed and their size is very small. After annealing, the peaks become sharper, and (202) and (020) peaks are separated.

Figure 2 shows the morphologies and the particle size distribution of powders prepared via SS (a), route M (b), and route MA (c). The particles in the SS powder are partially facetted; their average size estimated from TEM images is 130 nm. The crystallite size of M powder is 10 ~ 20nm; however, the particles are highly agglomerated, as shown in Fig. 2(b). In contrast, MA powder in Fig. 2(c) exhibits a narrow size distributionwith an average crystallite size of 10 ~ 20 nm, quite comparable with that of M powder, implying that the powder was deagglomerated during subsequent, annealing and wet-milling.

Table 1 lists the relative densities of different powder samples obtained after sintering at 1100°C for 12h in oxygen. The data obtained in previous investigations are also listed. The obtained relative density is the highest, 99.4%, for MA powder and the lowest, 82.7%, for M powder. It appears that the agglomeration of particles is detrimental to densification in KNN as in other systems.[19] It should also be noted that cracks were present in some M powder compacts after sintering. The crack formation may be due to the decomposition of unreacted residual carbonates in the powder, as observed previously in BaTiO3.[20,21] Additional annealing and deagglomeration by wet-milling can, therefore, ensure high sinterability of mechanochemically synthesized KNN powder. Full densification by pressureless sintering of KNN powder can easily be achieved.

The piezoelectric properties of different samples are also listed in Table 1 together with those obtained by pressureless sintering in previous investigations. The piezoelectric constant of the samples from MA powder is much higher than those from SS and M powder. It is also superior to those obtained in previous investigations. Egerton et al.[1] first reported a piezoelectric constant of 80 pC/N with 94.4 % of theoretical density. Guo et al.[4] show that KNN ceramics could exhibit 97 pC/N with 96.2 % of theoretical density through pressureless sintering at 1100 °C by adoption of cold isostatic pressing. Good piezoelectric properties were also obtained in a previous investigation by Zuo, el al.[22] They show that KNN ceramics could exhibit 107 pC/N with 98% of theoretical density through pressureless sintering at 1100 °C by adoption of additional attrition milling for 24h with calcined powder. In that case, the particle size after additional attrition milling was about 76nm. It should be noted that they used commercial oxide and carbonate powders to get KNN phases, just like in the present study. However they used the calcination process to get KNN phases and the additional attrition milling to reduce the particle size of the calcined powders. In the present study, we used the mechanochemical synthesis to get KNN phases without the calcinations process. Compared to previous reports, our optimized KNN shows superior piezoelectric properties.

IV. CONCLUSION

Optimization of pressureless sintering of KNN ceramics is described. Even pressureless sintering for dense KNN is required for its industrial application; however, to date, the optimization has not yet been established. But, to utilize its superior piezoelectric properties, pressure-assisted or field assisted sintering techniques have been attempted. To achieve pressureless sintering of KNN ceramics, we approached the synthesis of fine and uniform particles via the mecahnochemical technique. A particle size of 10-20 nm was successfully obtained and the powder compact was fully densified at 1100 °C for 12 h in oxygen and exhibited a piezoelectric constant of 119 pC/N and a coupling factor of 0.41.

ACKNOWLEDGMENTS

This work was supported by a grant from Korea Institute of Materials Science, a subsidiary branch of Korea Institute of Machinery and Materials.

REFERENCES

[1]L. Egerton, and D. M. Millon, " Piezoelectric and Dielectric Properties of Ceramics in the System Potassium Sodium Niobate," *J. Am. Ceram. Soc.*, **42**, 438-42 (1959).

[2]Y. Saito, H. Takao, T. Tani, T. Nonoyama, K. Takatori T.Honma, T. Nagaya, and N. Nakamura, "Lead-free piezoceramics," *Nature*, **423**, 84-7 (2004).

[3]E. Ringgaard, and T. Wurlitzer, "Lead-free piezoceramics based on alkali niobates," *J. Eur. Ceram. Soc.*, **25**, 2701-6 (2005).

[4]Y. Guo, K. I. Kakimoto, and H. Ohsato, "Phase transitional behavior and piezoelectric properties of $(Na_{0.5}K_{0.5})NbO_3$-$LiNbO_3$ ceramics," *Appl. Phys. Lett.*, **85**, 4121-3 (2004).

[5]V. Bobnar, J. Bernard, and M. Kosec, "Relaxorlike dielectric properties and history-dependent effects in the lead-free $Na_{0.5}K_{0.5}NbO_3$-$SiTiO_3$ ceramic system," *Appl. Phys. Lett.*, **85**, 994-6 (2004).

[6]K. Singh, L. Lingwal, S. C. Bhatt, and B. S. Semwal, "Dielectric properties of potassium sodium niobate mixed system," *Mat. Res. Bull.*, **36**, 2365-74 (2001).

[7]T. Takeuchi, T. Tani, and Y. Saito, "Piezoelectric properties of bismuth layer-structured ferroelectric ceramics with a preferred orientation processed by the reactive templated grain growth method," *Jpn. J. Appl. Phys.*, **38**, 5553-6 (1999).

[8]H. Birol, D. Damjanovic, and N. Setter, "Preparation and characterization of $(K_{0.5}Na_{0.5})NbO_3$," *J. Eur. Ceram. Soc.* , **26**, 861-6 (2006).

[9]J. F. Li, K. Wang, B. P. Zhang, and L. M. Zhang, "Ferroelectric and piezoelectric properties of fine-grained $Na_{0.5}K_{0.5}NbO_3$ lead-free piezoelectric ceramics prepared by spark plasma sintering," *J. Am. Ceram. Soc.*, **89**, 706-9 (2006).

[10]R. Wang, R. Xie, T. Sekiya, and Y. Shimojo, "Fabrication and characterization of potassium-sodium niobate piezoelectric ceramics by spark plasma sintring method," *Mat. Res. Bull.*, **39**, 1709-15 (2001).

[11]C. Herring, "The Physics of Powder Metallurgy," *W.E. Kingston (ed.), MaGraw-Hill, New York*, 143-79 (1951).

[12]C. Sun, X. Xing, J. Chen, J. Deng, L. Li, R. Yu, L. Aiao, and G. Liu, "Hydrothermal synthesis of single crystalline $(K,Na)NbO_3$ powders," *Eur. J. Inorg. Chem.*, 1884-8 (2007).

[13]Y. Shiratori, A. Magrez, and C. Pithan, "Particle size effect on the crystal structure of $K_{0.5}Na_{0.5}NbO_3$," *J. Eur. Ceram. Soc.*, **25** , 2075-9 (2005).

[14]Y. Shiratori, A. Magrez, and C. Pithan, "Microemulsion mediated synthesis of nanocrystalline $(K_x,Na_{1-x})NbO_3$ powders ," *Chem. Phys. Lett.*, **391** , 288-92 (2004).

[15]C. Pithan, Y. Shitratoni, J. Dornseiffer, FH. Haegel, A. Magrez, and R. Waser, "Microemulsion mediated synthesis of nanocrystalline $(K_{-x},Na_{1-x})NbO_3$ powders," *J. Crystal. Growth*, **280**, 191-200 (2005).

[16]J. Y Yun, J. H. Jeon, and S. J. L. Kang, "Synthesis of Sodium Niobate Powders by Mechanochemical Processing," *Mater. Trans.*, **49**, 2166-68 (2008)

[17]D. W. Baker, P. A. Thomas, N. Zhang and A. M. Glazer, "Structural study of $K_xNa_{1-x}NbO_3$ (KNN) for compositions in the range x=0.24-0.36," *Acta Cryst.*, **B65**, 22-8 (2009).

[18]H. Yang, Y. Lin, J. Zhu and F. Wang, "An efficient approach for direct synthesis of $K_{0.5}Na_{0.5}NbO_3$ powders," *Powder Tech.*, **196**, 233-6 (2009).

[19]W.H.Rhodes, "Agglomerate and particle size effects on sintering yttria stabilized zirconia," *J. Am. Ceram. Soc.*, **64**, 19-22 (1981).

[20]B. K. Yoon, E. Y. Chin and S.-J.L. Kang, "Densification during sintering of $BaTiO_3$ caused by the decomposition of residual $BaCO_3$," *J. Am. Ceram. Soc.*, **91**, 4121-4 (2008).

[21]J. Y. Yun, "Mechanochemical synthesis of $K_{0.5}Na_{0.5}NbO_3$ powders and its sintering behavior,"

Ph.D Thesis, Korea Advanced Institute of Science and Technology, (2009)
[22]R. Zuo, J. Rodel, R. Chen and L. Li, "Sintering and Electrical Properties of Lead-Free Na$_{0.5}$K$_{0.5}$NbO$_3$ Piezoelectric Ceramics," *J. Am. Ceram. Soc.*, **89**, 2010-5 (2006).

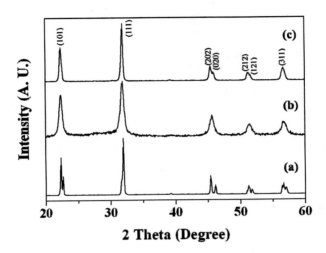

Figure 1. XRD patterns of KNN powders prepared by solid state (SS) reaction (a), mechanochemical synthesis without annealing (route M) (b), and mechanochemical synthesis with annealing (route MA) (c).

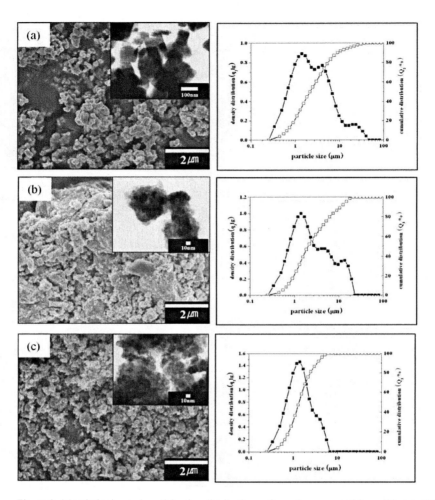

Figure 2. Morphologies and particle size distributions of powders prepared by solid state (SS) reaction (a), mechanochemical synthesis without annealing (route M) (b), and mechanochemical synthesis with annealing (route MA) (c).

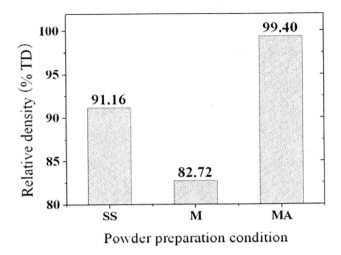

Figure 3. Sintered density of KNN prepared by solid state (SS) reaction, mechanochemical synthesis without annealing (route M), and mechanochemical synthesis with annealing (route MA).

Table 1. Comparison of properties between our work and other studies.

Properties	SS	M	MA	Egerton[1]	Guo[4]	Zuo[23]
Relative density (%)	91.2	82.7	99.4	94.4	96.2	98
Piezoelectric constant d$_{33}$ (pC/N)	83	~70	119	80	97	107
Coupling factor k$_p$ (%)	-	-	0.41	0.36	0.29	0.40
Dielectric constant at 1 kHz	-	-	419	290	402	580

SYNTHESIS OF POLYCRYSTALLINE Sr$_2$Fe$_{1+x}$Mo$_{1-x}$O$_6$ SAMPLES PRODUCED BY SOLID- STATE REACTION

Reginaldo Mondragón[1,2,3], Ricardo Morales[4], José Lemus-Ruiz[4], Oracio Navarro[1]

[1] Instituto de Investigaciones en Materiales, Universidad Nacional Autónoma de México, Apartado Postal 70-360, 04510 México D.F., México.
[2] Centro de Investigaciones en Materiales Avanzados, S.C., Chihuahua, Chih., México.
[3] Facultad de Ingeniería Química, Universidad Michoacana de San Nicolás de Hidalgo, Morelia, Mich., México
[4] Instituto de Investigaciones Metalúrgicas, Universidad Michoacana de San Nicolás de Hidalgo, Edif. U, CU, Apdo. Postal 888, C.P. 58000, Morelia, Mich., México

ABSTRACT

The main objective of this work was to study various aspects during the production of Sr$_2$Fe$_{1+x}$Mo$_{1-x}$O$_6$ samples with different x composition by solid-state reaction method. The synthesis consisted of three steps; ball milling, calcination, and reduction. Initially, stoichiometric combination of high purity powders of SrCO$_3$ (99.99%), Fe$_2$O$_3$ (99.98%), and MoO$_3$ (99.9%) were mixed using a high energy ball milled at 1800 rpm for 6 h. Then, 20 mg of powder were calcinated in dried air at 1100°C. Each calcinated sample was reduced at 1200°C under a flow of reducing gas mixture of Ar/5%H$_2$. Both, calcination and reduction experiments were carried out non-isothermally using thermogravimetric equipment which has an accuracy of 0.03 µg. The microstructural characterization was carried out by SEM, TEM, and XRD. The results showed the formation of a precursor SrMoO$_4$ phase in powders analyzed after mechanical ball milling process. On the other hand, analysis of samples calcinated for 2 h confirmed the existence of SrMoO$_4$ and Sr$_2$Fe$_2$O$_5$ phases which are precursors of the Sr$_2$FeMoO$_6$ compound. XRD analyses showed the formation of Sr$_2$ Fe$_{1+x}$ Mo$_{1-x}$ O$_6$ compound after the reduction process. The phase structure of the double perovskite was confirmed by Transmition Electron Microscopy.

INTRODUCTION

Double perovskite compounds have attracted intense research activities in both theoretical and applied areas of solid state physics and materials science due to their interesting structural, magnetic and electronic properties. While the concept of a crystalline solid as a perfect periodic structure is at the core of our understanding of a wide range of material properties, disorder is in reality ubiquitous, and is capable of influencing various properties drastically [1, 2]. Most theoretical treatments and experimental interpretations assume uncorrelated distribution of disordered sites, while it is known that physical properties may be strongly influenced by the exact nature of the distribution of disordered sites in a material. One of the most encouraging materials is the ordered double perovskite oxide materials of the form A$_2$BB'O$_6$ which consists of the alternative arrangement of ABO$_3$ and AB'O$_3$ perovskite units (A: alkaline-earth or rare-earth elements, B, B':transition-metal elements). The metallic and magnetic character depends primarily on the choice of B and B', but is also affected by the *local ordering* of these ions [3, 4]. Polycrystalline bulk samples have been extensively investigated for fundamental studies aimed to establish the structural and fundamental properties of the ideal crystal structural of the ordered double perovskite compound. Electronic structure calculations and optical spectroscopy investigations indicate that Sr$_2$FeMoO$_6$ is of half metallic nature [5]. Recently, Sr$_2$FeMoO$_6$ has attract much attention due its large room temperature magnetoresistance and high curie temperature of $T_c\sim410K$ higher than that of manganites, making it a potential candidate for spintronic device applications such as spin valves and magnetic tunnel junctions [6, 7]. Double perovskite Sr_2FeMoO_6 is a kind of spintronics material exhibiting large negative magnetoresistance at high temperatures and low magnetic fields presumes a perfect alternate occupancy of the Fe and Mo ions along the three cube axes, giving rise to the effect of doubling the basic unit perovskite cell. However, the presence of the Fe/Mo disorder, for example, creation of imperfections by interchanging Fe and Mo ions positions in a perfectly ordered structure. Sr_2FeMoO_6 exhibits unique trends both in its magnetic as well as magnetoresistance behavior where the degree of this disorder is found to play a critical role. Moreover, the electronic structure of Sr_2FeMoO_6 has also been recently shown to get influenced by Fe/Mo cation disorder. This can happen because the valence difference and ionic radius difference between Fe^{3+} and Mo^{5+} fall in the region where ordered and disordered double perovskites can coexist [8, 9].

Samples of Sr_2FeMoO_6 are usually prepared by the solid-state reaction or sol–gel method. However, it is prone to form anti-site defects (ASD), anti-phase domains (APD) and intrinsic disorder under certain conditions [10-12]. The insulating matrix of the anti-phase defect and intrinsic disorder inside the Sr_2FeMoO_6 nanoparticles can reduce the eddy current loss and make it possible to be used as a microwave absorption material. L. Xi et al [13] produced Sr_2FeMoO_6 nanoparticles by the sol–gel method at different sintering temperatures, which are the key factors to get the ideal half-metallic properties of Sr_2FeMoO_6. On the other hand, E. Burzo et al. (2007)[14] and E. Burzo et al. (2011) [15] prepared samples of Sr_2FeMoO_6 by standard solid-state reaction of $SrCO_3$, Fe_2O_3 and MoO_3 at 1300°C in Ar/H_2 atmosphere. They shown that the nonuniform composition inside the grains and that at grain boundaries influence the magnetoresistive properties and the grains are more homogeneous when sintering time increased. The two important parameters for possible applications of double perovskite Sr_2FeMoO_6 are its Curie temperature (T_C) and the magnetoresistance. Therefore, in order to enhance the T_C of Sr_2FeMoO_6 injection of electrons into the conduction band by appropriate doping is a natural strategy. Q. Zhang et al. [16] investigated the crystal structure, magnetic and electrical-transport properties of the rare-earth-doped compounds preparing the samples by standard solid-state reaction of stoichiometric mixed powders. They found that the injection of doping electrons into the Mo orbital reduces the charge difference between the Fe and the Mo ions, leading to a decrease of the degree of cationic ordering of the doped compounds. In this work, we report the synthesis and characterization of double perovskite Sr_2FeMoO_6.

EXPERIMENTAL PROCEDURE

The starting materials were high purity powders of: $SrCO_3$ (Aldrich, 99.9%), Fe_2O_3 (Aldrich, 99.98%), and MoO_3 (Aldrich, 99.9%). Prior to weighing, the powders were dried at 100°C for 16 h, then, samples with stoichiometric proportions were mixed using an agate mortar. In order to obtain the $SrMoO_4$ precursor phase, the samples were milled using a high energy ball milling at 1800 rpm in an 8000 M Spex Mixer for 6. Both calcination and reduction experiments were carried out using a non-isothermal cicle until 1100°C in a thermogravimetric equipment (SETARAM, Setsys Evolution) with a precision 0.03 μm. The weight change during experiments was recorded at 1 s intervals. Twenty mg of powder were used inside an alumina crucible suspended from one arm of the balance. The temperature of the furnace was measured by a Pt-10%Rh (S type) thermocouple placed below the crucible. The working gases flows were individually regulated using mass flow controllers. Calcinations experiments were carried out using dry air at constant flow following a heating rate of 15°C/min up to 650°C followed by a ramp up to 1100°C at a heating rate of 5°C/min. The experiments were finished by cooling down the furnace until room temperature keeping the dry air atmosphere. On the other hand, the reduction experiments were carried out using calcinated samples in a mixture of $Ar/5\%H_2$ atmosphere at constant flow. The reduction samples were heated from room temperature up to 1100°C at heating rate of 5°C/min. The reduction gas mixture was kept throughout the experiment in order to avoid oxidation of the reduced sample during the cooling process of the furnace.

Microestructural characterization for phase identification behavior, during the different steps of the process, were carried out using X-ray diffraction (XRD), scanning electron microscope (SEM) and transmission electron microscopy (TEM). Initially, the purity of all materials was verified, and the resulting powder of ball milling, calcinations and reduction experiments were characterized for phase's identification using a D-5000 Siemens X-ray diffractometer with CuK_α radiation. The morphology examination of the powders was performed using both the secondary electron (SEI) and backscattered modes (BSI) in a JEOL JSM-6400 scanning electron microscope. Further compositional analysis was performed using electron probe microanalysis with wavelength dispersive system (WDS). A High resolution TECNAI F20 transmission electron microscope was employed to study particle microstructure.

RESULTS AND DISCUSSION

The process starts when the high purity MoO_3, $SrCo_3$, and Fe_2O_3 powders were mixed in stoichiometric amounts using an agate mortar and then ball milled for 6 h. The starting powders materials were characterized to determine different morphology and particle size distribution of the materials. For example, MoO_3 powder presented laminar shape, however $SrCO_3$ and Fe_2O_3 powders had a regular shape and smaller particle size than MoO_3 (Fig 1a). The effect of the ball milling on the reduction of particle size can be observed in Fig 1b. It can be observed that the powder mixture subjected to mechanical milling showed a particle size reduction, as well as a homogeneous morphological distribution.

a)   b)

Figure 1. Scanning electron micrographs of powders a) before and b) after ball milling.

On the other hand, the objective of using ball milling was to promote the formation of precursor phase $SrMoO_4$ while the unreacted specie, Fe_2O_3, is both energy-activated and mixed thoroughly. X-ray analysis carried out in the ball milled powders shows the presence of peaks corresponding to $SrMoO_4$ which is a precursor of the Sr_2FeMoO_6 compound. Fig 2 shows the X- ray diffraction patterns of the sample milled for 6 h. It can be confirmed that several structural changes occurred during milling process lead to the formation of the $SrMoO_4$ phase, which is clearly identified by several peaks, including the three stronger broadened diffraction peaks at 2θ angles of 27.88, 45.42, and 56.38°, respectively. These results indicate that the mechanical milling induced the $SrMoO_4$ phase formation.

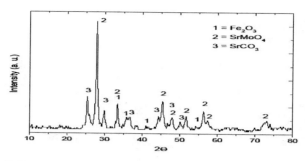

Figure 2. X-ray spectrum of powder after high energy mechanical alloyed for 6 h.

The second step, corresponding to calcinations, of the synthesis route had the purpose of forming the second precursor phase, $Sr_2Fe_2O_5$. According to the XRD results of Fig 2, the initial phases before the calcinations step are Fe_2O_3, $SrMoO_4$, and $SrCO_3$. During this step the sample presented a total weight loss of 12.1%, which can be associated to different factors of the decomposition of strontium carbonated. Arvanitidis et al. [17] reported that the decomposition of pure $SrCO_3$ under argon atmosphere occurs at about 700°C. Fig 3 shows the X-ray spectrum for the sample after calcinations process. It can only be observed, from Fig 3, the presence of $SrMoO_4$ and $Sr_2Fe_2O_3$ phases.

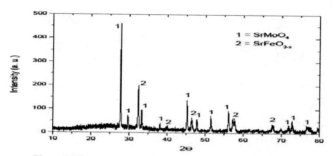

Figure 3. X-ray spectrum of powder after calcinations process.

The third processing step was to promote the reaction of the precursor phases, produced during high energy and calcinations steps, under controlled reducing atmosphere. Fig 4 shows the reduction curve under a constant heating rate. The reduction path seems to be dependent upon temperature and occurs in two main stages, namely, from 300-600°C and above 600°C. The oxygen atoms in excess are likely to be removed from both $SrMoO_4$ and $Sr_2Fe_2O_3$ particles at slow rate. $SrMoO_4$ should reduce first followed by $Sr_2Fe_2O_3$ since MoO_3 and Fe_2O_3 pure powders are reduced by hydrogen in the same order temperature-wise [18]. This may explain the two reduction stages seen in Fig 4.

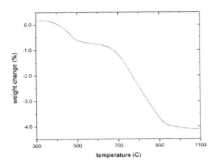

Figure 4. Not-isothermal reduction curve of precursor powders in $Ar/5\%H_2$.

It is likely that the partial reduction of $SrMoO_4$ would promote the diffusion of Mo ions into the $Sr_2Fe_2O_3$ structure as temperatures increases and partial reduction proceeds. This evidence agrees with T.T. Fang *et al.* [19] in that Sr_2FeMoO_6 is formed via the substitution of Mo for Fe in $Sr_2Fe_2O_3$. X-ray analyses performed on the reduced samples revealed that the diffraction patterns fit best the $Sr_2Fe_{1.3}Mo_{0.7}O_6$ phase agree with *Pattern: 01-072-6390 of ICSD Collection Code: 96235*, as shown in Fig 5.

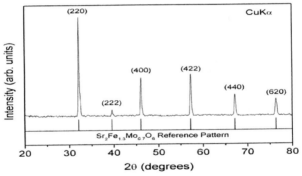

Figure 5. X-ray spectrum of powder after reduction process.

The phase analyses of the powders produced by no-isothermal reductions experiments until 1000°C were also examined by transmission electron microscopy. Fig 6a shows a high resolution TEM micrograph showing a nano-crystal of non-stoichiometric perovskite $Sr_2Fe_{1.3}Mo_{0.7}O_6$ phase and its corresponding fast Fourier transform (FFT) pattern. The image shows enlarge morphology of 8 nm wide and 30 nm long. Typical defects corresponding to phase transformations are not observed, this suggests that the double perovskite phase can be stable. The fast Fourier transform (FFT) shows a diffraction pattern which corresponds to the (220) lattice plane of the crystalline double perovskite structure $Sr_2Fe_{1.3}Mo_{0.7}O_6$. Fig 6c shows the filtered image of the structure, the interplanar spacing of 0.278 nm corresponds to that of the (220) plane of $Sr_2Fe_{1.3}Mo_{0.7}O_6$ phase. These results are in agreement with the previous XRD results.

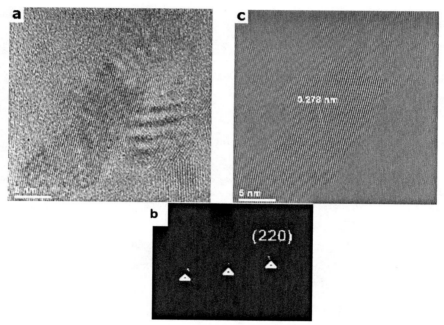

Figure 6. a) A high resolution TEM micrograph of the double perovskite $Sr_2Fe_{1.3}Mo_{0.7}O_6$ non-isothermally reduced until 1000°C in 5%H_2, b) diffraction pattern and c) filtered processed image.

CONCLUSIONS

Mechanical ball milling renders homogeneous distribution of particles with reduction in size as well as the partial formation of a precursor phase, Sr_2MoO_4. Thus, by mechanically activating precursor powders and controlling the process parameters of the gas-solid reaction steps, i.e. calcination and reduction processes, the kinetics formation of double perovskites structures, in particular $Sr_2Fe_{1.3}Mo_{0.7}O_6$ can be simplified. Consequently, both processing parameters, temperature and time, are significantly reduced. In this work, the double perovskite phase $Sr_2Fe_{1.3}Mo_{0.7}O_6$ was synthesized by a three steps process using a maximum temperature of 1000°C.

ACKNOWLEDGMENT

The authors must want to acknowledge support by PAPIIT IN108710 from UNAM, CONACYT project No. 131589, and Universidad Michoacana de San Nicolás de Hidalgo.

REFERENCES

[1] S.E.A. Yousif and O.A. Yassin, The Electronic and Magnetic Propeties of Sr_2MnNbO_6, Sr_2FeMoO_6 and Sr_2NiRuO_6 Double Perovskites: An LSDA+U+SOC Study, *J. Alloys and Compounds*, **506** 456-

460 (2010).

[2] C. Meneghini, S. Ray, F. Liscio, F. Bardelli, S. Mobilio, and D.D. Sarma, Nature of "Disorder" in the Ordered Double Perovskite Sr_2FeMoO_6, *Physical Review Letters*, **103** 046403-1-046403-4 (2009).

[3] P. Sanyal, S. Tarat, and P. Majumdar, Structural Ordening and Antisite Defect Formation in Double Perovskites, *Eur. Phys. J. B*, **65** 39-47 (2008).

[4] K. Yoshida, S. Ikeuchi, H. Shimizu, S. Okayasu, and T. Suzuki, Strong Correlation Among Structural, Electronic, and Magnetic Properties of $Sr_2Fe_{1+x}Mo_{1-x}O_6$ ($0 \leq X \leq 1$), *J. Phys. Soc. Jpan.*, **80** 044716-1-044716-4 (2011).

[5] K.I. Kobayashi, T. Kimula, H. Sawada, K. Terakura, Y. Tokura, Room-Temperature Magnetoresistance in an Oxide Material with an Ordered Double-Perovskite Structure, *Nature*, **395** 677-680 (1998).

[6] F. Azizi, A. Kahoul, and A. Azizi, Effect of La Doping on the Electrochemical Activity of Double Perovskite Oxide Sr_2FeMoO_2 in Alkaline Medium, *J. Alloys and Compounds*, **484** 3259-3266 (2010).

[7] D. Kumar and D. Kaur, Substrate-dependent Structural and Magnetic Properties of Sr_2FeMoO_6 Nonostructured Double Perovskite Thin Films, *Physica B*, **405** 3259-3266 (2010).

[8] Q. Zhang, G.H. Rao, Q. Huang, X.M. Feng, Z.W. Ouyang, G.Y. Liu, B.H. Toby, and J.K. Liang, Selective Substitution of Vanadium for Molybdenum in $Sr_2(Fe_{1-x}V_x)MoO_6$ Double Perovskites, *J. Solid State Chemical*, **179** 2458-2465 (2006).

[9] B. Manoun, S. Benmokhtar, L. Bih, M. Azrour, A. Ezzahi, A. Ider, M. Azdouz, H. Annersten, and P. lazor, Synthesis, Structure, and High Temperature Mössbauer and Raman Spectroscopy Studies of $Ba_{1.6}Sr_{1.4}Fe_2WO_9$ Double Perovskite, *J. Alloys and Compounds*, **509** 66-71 (2011).

[10] Z. Wang, Y. Tian, and Y. Li, Direct CH_4 Fuel Cell Using Sr_2FeMoO_6 as an Anode Material, *J. Powder Sources*, **196** 6104-6109 (2011).

[11] L.M. Torres, I. Juárez, X.L. García, A. Cruz, Desarrollo de Semiconductores con Estructuras Tipo Perovskitas para Purificar el Agua Mediante Oxidaciones Avanzadas, *Ciencia UANL*, **XII**, 4 376-388 (2010).

[12] S.E. Jacobo, Novel Method of Synthesis for Double-perovskite Sr_2FeMoO_6, *J. Materials Science*, **40** 417-421 (2005).

[13] L. Xi, X.N. Shi, Z. Wang, Y.L. Zuo, and J.H. Du, Microwave Absorption Properties of Sr_2FeMoO_6 Nanoparticles, *Physica B*, **406** 2168-2171 (2011).

[14] E. Burzo, I. Balasz, S. Constantinescu, and I.G. Deac, Grain Boundary Effects in Highly Ordered Sr_2FeMoO_6, *J. Magnetics Materials*, **316** c741-c744 (2007).

[15] E. Burzo, I. Balasz, M. Valeanu, and I.G. Pop, The Effects of Thermal Treatment on the Physical Properties of $Sr_2FeMo_{1-x}M_xO_6$ Perovskite with M=W, Ta and $X \leq 0.3$, *J. Alloys and Compounds*, **509** 105-113 (2011).

[16] Q. Zhang, G.H. Rao, Y.G. Xiao, H.Z. Dong, G.Y. Liu, Y. Zhang, and J.K. Liang, Crystal Structure, Magnetic and Electrical-transport Properties of Rare-earth-doped Sr_2FeMoO_6, *Physica B*, **381** 233-238 (2006).

[17] I. Arvanitidis, D. Sichen, H.Y. Sohn, and S. Seetharaman, The Intrinsic Thermal Decomposition Kinetics of $SrCO_3$ by a Nonisothermal Technique, *Metall. Mater. Trans. B*, **28** 1063-1068 (1997).

[18] P. Pourghahramani and E. Forssberg, Reduction Kinetics of Mechanically Activated Hematite Concentrate with Hydrogen Gas Using Nonisothermal Methods, *Thermochimica Acta*, **454** 69-77 (2007).

[19] T.T. Fang, M.S. Wu, and T.F. Ko, On the Formation of Double Perovskite Sr_2FeMoO_6, *J. Mater. Sci. Lett.*, **20** 1609-1610 (2001).

Interfacial Reaction
and Sintering

EFFECTS OF CHEMICOPHYSICAL PROPERTIES OF CARBON ON BLOATING CHARACTERISTICS OF ARTIFICIAL LIGHTWEIGHT AGGREGATES USING COAL ASH

Shin-hyu Kang, Ki-gang Lee, Yoo-taek Kim, and Seung-gu Kang
Dept. of Advanced Materials Science and Engineering, Kyonggi University
Suwon, Gyeonggi, Republic of Korea

ABSTRACT

The purpose of this study was to figure out the impacts of Carbon types and dosages to foaming when producing artificial lightweight aggregates by utilization of recycled resources such as bottom-ash, reject-ash and dredged-soil. In order to figure out chemical characteristics of raw materials, XRD and XRF analyses were performed, and in order to observe foaming characteristics according to differences in reaction constants of Carbon, DTF analyses of C/B(Carbon Black) and A/C(Activated Carbon) were performed. 30wt% of dredged-soil and 70wt% of each calcined bottom-ash-A, bottom-ash-B and reject-ash were mixed, then the amount of Fe_2O_3 was added to be fixed at 11wt%, and then Carbon was added. As molded aggregates were sintered by high speed sintering in intervals of 40°C from 1060 °C to 1180°C, specific gravity and water absorption were measured. As a result, the artificial lightweight aggregate with Carbon of 3-6Vol% showed the lowest specific gravity, and it was identified that the more Carbon Vol% increases, the more specific gravity increases. In addition, as C/B showed rapid burning speed value, the addition of even a small amount helped in light-weighing aggregates, and it did not interfere with surface formation of aggregates, which showed lower water absorption compared to that of A/C, which has relatively low reaction constant.

INTRODUCTION

In domestic thermoelectric power plants, about 9 million tons of coal-ash is generated a year, and except for some Fly ash, Reject-ash and Bottom-ash are difficult to recycle, thus causing depends on landfill. The generated volume of Bottom-ash reaches 1.6 million tons a year, which is 15-20% of total coal-ash, and although its characteristics and grading are not uniform, the reality is relying on landfills.

Recently, various studies on recycling this bottom-ash are underway, and the most actively studied aspect is recycling as artificial lightweight aggregates. Aggregate serves an important function in construction, and as demand of such aggregate increases, it causes serious environmental damage such as digging hills to collect aggregates. To solve this problem, coal-ash is to be recycled as artificial lightweight aggregate, which prevents environmental damage, and waste is to be used as recycled material instead of being dumped in landfills. In addition, artificial lightweight aggregate has soundproofing and thermal isolation features, causing it to be highlighted as a multi-purpose construction material.

Riley identified the foaming mechanism by adding silica and alumina to non-foaming clay, which established a Bloating zone, However, he did not consider the generation of foaming gas caused by Carbon and molten phase by Fe. Thai et al. presented the findings on the characteristics of lightweight aggregate generated by sintering from incinerated material of waste sludge primarily composed of SiO_2-Al_2O_3-flux, however asserted that it was different from the existing foaming ternary system and predicted there was another factor affecting foaming. For foaming gas, Jackson asserted that foaming caused by foaming gas was generated during decomposition of iron oxide contained in clay, and Dettmer asserted that foaming was generated as Fe_2O_3 or other gas contained in raw materials was dissolved and supersaturated in liquid to generate the core of glass structure. However he could not prove the foaming effect was caused by Carbon. Park et al. asserted that the more Carbon dosage calcination temperature increased, the more area of black core formation and the presence of critical content of Carbon increased. However he could not prove the light-weighting effect on aggregate.

In this study, it was intended to verify the impact of the combustion speed and content of Carbon on light-weighting in utilization as an artificial lightweight aggregate by using unburned carbon contained in coal-ash, and to understand the foaming mechanism.

2. EXPERIMENTAL

In this study, coal-ash and dredged-soil remaining from coal combusted in the Y thermoelectric power plant were used as raw materials. Pin mill was used to grind bottom-ash and dredged-soil to less than 100μm to produce an artificial lightweight aggregate. Ground raw materials characteristics were analyzed by XRF, and the chemical composition of each raw material was as shown in Table 1. Two types of bottom ash used for raw materials because of the different content of unburned Carbon.

Table 1. Chemical Composition of Coal-ash and Dredged Soil

	Bottom-ash-A	Bottom-ash-B	Reject-ash	Dredged Soil
Ig. loss	4.07	1.04	3.52	4.04
SiO_2	45.62	61.00	51.33	71.00
Al_2O_3	18.59	25.40	21.61	14.24
Fe_2O_3	8.07	4.14	5.33	3.78
CaO	2.17	1.00	4.28	0.78
MgO	0.78	0.94	1.09	0.18
Na_2O	0.18	0.08	0.82	2.49
K_2O	0.51	3.23	0.99	2.67
TiO_2	1.33	0.86	1.16	0.79
P_2O_5	0.24	0.12	0.59	0.03
MnO	0.05	0.03	0.15	0.00
SO_3	0.00	0.43	0.35	0.00
Cr_2O_3	0.01	0.02	0.00	0.00
ZrO_2	0.33	0.00	0.00	0.00
C	18.05	1.74	8.68	0.0
Total	100	100	100	100

In previous studies, it was identified that the most effective content ratio of Fe_2O_3 for aggregate foaming was about 11wt%. Accordingly, reagent of Fe_2O_3 (Kanto Chemical Co., Inc, 95.0%, Japan) was used to fix content of Fe_2O_3 of all the mixing ratios into 11wt%. Before adding addition agent, the mixing ratio of coal-ash and dredged-soil was fixed to 7:3, the moldable condition in the Pelletizer, and the condition of maximizing recycling of coal-ash. In order to compare foaming characteristics of artificial lightweight aggregates according to Carbon types and dosages, Coal-ash was calcined to completely remove unburned carbon, and then used. For added Carbon, Carbon Black(C/B) and

Activated Carbon (A/C) were used, respective mixing ratios and calcination temperatures were as shown in Table 2.

Throughout the mixing process, prepared raw materials were used to mold the artificial lightweight aggregate into about a 10mm ball. The molded aggregate was dried at 105 °C for 24 hours, and calcined in intervals of 40 °C ranging from 1060°C to 1180°C. When inserting the aggregate, the calcination method used was high speed sintering (heating rate : 20°C/min) which keeps the object at the highest temperature for 10 minutes and discharges the object.

The produced artificial lightweight aggregates were measured for bulk specific gravity and water absorption in accordance with KS L 3114 standard method. In addition, in order to identify impact of combustion speed and reaction constant of Carbon used for foaming, DTG (Thermal Gravimetric Analysis, TA Instrument, SDT Q600, U.S.A.) analysis was performed.

Table 2. Batch & Sintering Temperature for Artificial lightweight aggregates

Raw materials		Added Materials (g/100g of raw materials)		Sintering Temp. (°C)
Calcined B/A-A	D/S	Fe_2O_3	Carbon Black	
			0, 0.5, 1, 1.5, 2, 3, 4, 5	
70 wt%		4	Activated Carbon	
			0, 1, 2, 3, 4, 5, 7	
Calcined B/A-B			Carbon Black	
	30 wt%	7	0, 1, 2, 3, 4, 5	1060, 1100, 1140, 1180 for each batch
70 wt%			Activated Carbon	
			0, 1, 2, 3, 4, 5	
Calcined R/A			Carbon Black	
		8	0, 1, 2, 3, 4, 5	
70 wt%			Activated Carbon	
			0, 1, 2, 3, 4, 5	

3. RESULT AND DISCUSSION

3.1. Raw materials

XRD results for raw materials used in this experiment were as shown in Fig. 1. In raw materials of (a), (b) and (c), RO_2 group states such as Quartz and Sillimanite were mainly shown, and for Fe, it was identified that it existed in the compounds such as Hercynite rather than Hematite. In addition, groups of R_2O_3 and RO/R_2O such as Albite, Anorthite and Holloysite were contained in dredged-soil (d). Therefore, because the combination of coal-ash and dredged-soil could satisfy whiteware body, it was determined that it could be used as raw materials of aggregates.

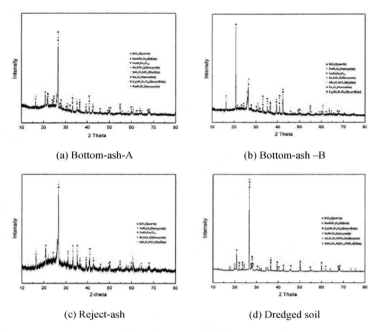

(a) Bottom-ash-A (b) Bottom-ash –B

(c) Reject-ash (d) Dredged soil

Fig. 1. XRD patterns of raw materials, (a) bottom-ash-A, (b) bottom-as-B, (c) Reject-ash and (d) dredged soil

3.2. Impact of Carbon Vol% and Fe2O3 Dosage

The artificial lightweight aggregate is that light-weighting occurs by foaming, and foaming mechanism is the surface formation that enables foaming gas generation and collection of generated gas. Foaming by Fe and Carbon is as follows.

$$C + O_2 \rightarrow CO_2 \qquad\qquad -- (1)$$
$$C + 1/2O_2 \rightarrow CO \qquad\qquad -- (2)$$
$$C + CO_2 \rightarrow 2CO \qquad\qquad -- (3)$$
$$2Fe_2O_3 \rightarrow 4FeO + O_2 \qquad\qquad -- (4)$$
$$3Fe_2O_3 \rightarrow 2FeO \cdot Fe_2O_3 + 1/2O_2 \qquad\qquad -- (5)$$

Equations of (1), (2) and (3) are that Carbon meets O_2 during calcination, then forms foaming gas, and then expands inside. Equations of (4) and (5) are the reducing reaction occurring in the high temperatures greater than 1000°C, which is the foaming mechanism of the artificial lightweight aggregate. Such foaming gas generating reactions have a close relationship with the behavior of the surface collectable of foaming gas.

Vol%'s according to Carbon types and dosages were as shown in Table 3. Dosage of C/B is smaller than that of A/C, however when converting to Vol%, they have the same values, which results from the difference in weights per volume of both substances, and means C/B has a bigger comparative area than A/C.

Table 3. Carbon Vol% of each batch

Added(g) Vol%	1.0	2.0	3.0	4.0	5.0
C/B	3.0	6.0	9.0	12.0	15.0
A/C	1.5	3.0	4.5	6.0	9.0

At calcination temperature of 1180°C, the specific gravity values according to each Carbon Vol% by raw materials were as shown in Fig. 2. All three of the raw materials showed high specific gravity at low Carbon Vol%, and the lowest specific gravity between 3-6 Vol%. Again, the more Vol% increased, the more specific gravity increased. It was considered for the composition with low Carbon Vol% that foaming of aggregate did not occur due to the the foaming gas. CO_2 was not sufficiently generated during sintering, causing light-weighting of aggregate to be difficult. In addition, the specific gravity for the composition with high Carbon Vol% was increasing because carbon was present as inclusion at surface sintering - because carbon existing on the aggregate surface was not completely oxidized - and foaming gas inside of aggregate was not collected – because surface densification rate decreased.

At Fig. 2 (a), Carbon Vol% of the lowest specific gravity for aggregate with raw materials of B/A-A was shown as 3Vol%. It could be noticed that Carbon Vol% indicating the lowest specific gravity of aggregate with raw materials of B/A-B and R/A was 6Vol% Fig. 2 (b,c). It was due to the partial pressure difference between P_{CO} and P_{CO2} according to P_{O2}, and based on equation (2), each P_{CO} and P_{CO2} at 1400K are defined in the following equations:

$$\log P_{CO} = 1/2 \log P_{O2} + 8.711 \qquad \text{-- (6)}$$
$$\log P_{CO2} = \log P_{O2} + 14.785 \qquad \text{-- (7)}$$

According to the above equations, both P_{CO} and P_{CO2} have the same value when P_{O2} is $10^{-12.145}$ atm. Based on this point, if P_{O2} decreases, it is advantageous to foaming because CO gas is discharged in the stable state. However, if P_{O2} increases, it is disadvantageous to foaming because CO_2 gas is in the stable state. In other words, it means that the generation of a tiny amount of O_2 prevents foaming of the aggregate.

Because Fe contained in coal-ash is present already in the form of spinel, it does not generate O_2 during calcination. However, if Fe_2O_3 is used as an addition agent, added Fe_2O_3 becomes Fe_3O_4 and discharges O_2. At this moment, because discharged O_2 increases P_{O2}, P_{CO2} increases and this prevents foaming of the aggregate. Therefore, in order to make P_{CO} conditions sufficient for foaming of the aggregate by decreasing P_{O2}, a greater amount of Carbon is needed for the composition using B/A-B and R/A, rather than for the composition using B/A-A, which was considered that Carbon Vol% showed the lowest specific gravity at 6Vol%.

In other words, it can be noticed that Vol% of Carbon should not be too low for foaming of the aggregate, and there is an appropriate Carbon Vol% that helps foaming of the aggregate and does not prevent surface densification.

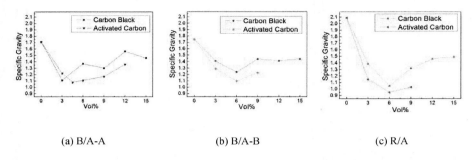

(a) B/A-A (b) B/A-B (c) R/A

Fig. 2. Bulk specific gravity of aggregates at 1180°C, (a) B/A-A, (b) B/A-B and (c) R/A

3.3. Reaction Rate and Water absorption According to Dosages of C/B and A/C

In order to identify the impact of foaming of the aggregate according to the reaction rate difference between C/B and A/C, DTG analysis was performed and the results were as illustrated in Fig. 3. Heating rates are 5°C, 10°C, 15°C, 20°C per minute, respectively. Based on the DTG results, reaction constant k was calculated and listed in Table 4. As shown in Fig. 3, the case of C/B addition showed a rapid response regardless of the rate of temperature increase, compared to the case of A/C addition. Results on Table 4, C/B shows rapid response because the fast reaction rate that has the higher A (frequency) value than A/C and wider surface area than A/C. In addition, it was identified that the reaction constant of C/B was 7.098E+7, which was much higher than 3.072 of A/C. By this, in dosage, it was noticed that C/B is more effective in foaming of the artificial lightweight aggregate than A/C.

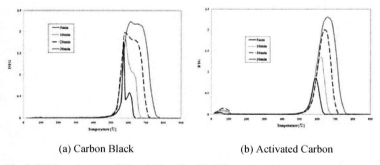

(a) Carbon Black (b) Activated Carbon

Fig. 3. DTG analysis (a) C/B and (b) A/C with different heating rate

Table 4. Rate constant for C/B and A/C

	E (kJ/mole)	A ($1min^{-1}$)	k ($1s^{-1}$)
Carbon Black	222.97	3.62E+19	7.098E+7
Activated Carbon	180.68	2.05E+10	3.072

As shown in Fig. 4, water absorption values of C/B and A/C were compared according to temperatures by each raw material at Carbon 6Vol%. It shows the tendency that water absorption of C/B is maintained or decreased as temperature increases; however that of A/C is increased. This is caused by the difference in reaction constants. It was considered that for C/B with high reaction constant, during sintering, gas was generated, expanded inside and discharged outside, which did not prevent surface formation, and thus the size of open blow holes was very small. However, for A/C with a low reaction constant, it took a long time for gas to be generated and discharged, which prevented surface densification of the aggregate, and thus it formed surface blow holes of bigger sizes.

It was judged that such results, according to Carbon types and content amount, would help further in the light-weighting of aggregates for construction materials and effective recycling of various waste substances such as coal-ash, and furthermore in the development of a super lightweight aggregate.

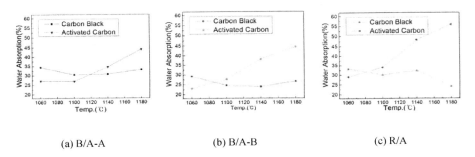

(a) B/A-A (b) B/A-B (c) R/A

Fig. 4. Water absorption of aggregates at Carbon 6Vol%, (a) B/A-A, (b) B/A-B and (c) R/A

4. CONCLUSION

1) As the result of XRD analysis, coal-ash is mainly composed of RO_2 group and dredged-soil contains R_2O_3 and RO/R_2O groups, which can be used to create a triaxial whiteware body.

2) For light-weighting of aggregates: light weighting was difficult because low Carbon Vol% did not generate enough foaming gas and high Carbon Vol% prevented surface densification due to surplus Carbon. The best Carbon Vol% for light-weighting of aggregate was 3.00-6.00Vol%.

3) If Fe_2O_3 is used as addition agent for foaming, it is necessary that P_{CO} be in a stable state, and thus to accomplish foaming, a greater Carbon amount is needed than in compositions where Fe_2O_3 is added relatively less. .

4) As the result of DTG analysis, C/B (7.098E+7) showed a higher reaction constant than A/C (3.072).

For C/B with a fast combustion rate, because gas was discharged without preventing surface formation after foaming the aggregate, the water absorption of aggregate was low. However, for A/C, because foaming gas prevented surface formation and thus expanded the surface blow holes, it showed high water absorption.

ACKNOWLEDGEMENTS

We appreciate the CERAGREEN of Korea for providing financial support for this study.

REFERENCES

[1]C. M. Riley, Relation of Chemical Properties to the Bloating of Clays, *J. Am. Ceram. Soc.,* 34(4), 121-8 (1951)

[2]C. C. Tsai et al., Effect of SiO_2-Al_2O_3-flux Ration Change on the Bloating Characteristics of Lightweight Aggregates Material Produced from Recycled Sewage Sludge, *J. Hazardous Materials B,* 134, 87-93 (2006)

[3]T. E. Jackson, Changes in Color of Clays on Ignition of Clayware Kilns, *Transactions of the Am. Ceram. Soc.,* 37-43 (1903)

[4]F. Dettmer, Keram Z., 12(58), 373-82 (1961)

[5]J. Y. Park et al., The Mechanism of Black Core Formation, *J. Kor. Crystal Growth and Crystal Tjechnology,* 15(5), 208-215 (2005)

[6]S. H. Kang et al., Effect on the Bloating and Physical Properties of Artificial Lightweight Aggregates with Fe, *J. Kor. Ceram. Soc.,* (in press)

[7]V. Beltran et al., Formation of the Black Core During the Firing of Floor and Wall Tiles, Interceram, 3, 15-21 (1988)

[8]W. D. Kingery et al., Physical Ceramics, Wiley (1997)

[9]D. R Stull and H. Prophet, JANEF Thermochemical Table; Second Edition, NSRDS (1971)

SINTERING OF SILICON, EFFECT OF THE SAMPLE SIZE ON SILICA REDUCTION KINETICS AND DENSIFICATION

J.M. Lebrun,* J.M. Missiaen, C. Pascal
Laboratoire de Science et Ingénierie des Matériaux et Procédés, SIMaP, Grenoble INP-CNRS-UJF, Domaine Universitaire, BP 75, F-38402 Saint-Martin d'Hères, France

*Corresponding author: jean-marie.lebrun@simap.grenoble-inp.fr

ABSTRACT
    Sintering of fine silicon particles covered with a silica layer of nanometric thickness was studied under low water vapor pressure. The role of the silica layer on sintering kinetics of silicon was reviewed. The effect of the powder compact size on silica reduction kinetics, microstructure evolution and final density was studied. The results were compared to the prediction of a previous model proposed for silica reduction kinetics. A larger compact leads to the retardation of silica reduction and to the enhancement of densification.

INTRODUCTION
    Silicon is largely available on earth, but standard materials processing for photovoltaic applications requires crystallization of silicon ingots of high purity obtained through high energy consuming processes. Ingot cutting is responsible for a large material waste and leads to expensive production costs. Sintering of near net shape silicon wafers is thus an important issue.
    Previous works showed that densification of silicon is not favored because of significant grain coarsening occurring in the early stage of sintering. Depending on the authors, the coarsening mechanism could be vapor transport [1, 2, 3] or surface transport [4, 5]. From an analysis with Herring's scaling law, Greskovich and Rosolowski [1] concluded that a vapor transport mechanism is dominant. Shaw and Heuer [2] observed a non-uniform microstructure over the compacts with a relatively dense inner core surrounded by a surface region of coarse porosity. The coarse surface region was related to volatilization and condensation of silicon monoxide ($SiO_{(g)}$) occurring through the reduction of the native silica layer covering the silicon particles. Möller and Welsh [3] observed the same non-uniform microstructure. The measured activation energy for densification in the dense inner core (500 kJ mol^{-1}) was close to silicon lattice self-diffusion activation energy (476 kJ mol^{-1}) [6], leading the authors to conclude that lattice diffusion overcomes vapor transport mechanisms for particle diameters less than 100 nm. These conclusions are not consistent with grain boundary grooving experiments realized by Robertson [4] and Coblenz [5]. Surface diffusion was found to be the dominant mechanism in both studies and the diffusion coefficient for this mechanism was estimated. Neck growth rate measurements on polycrystalline spheres of 150 to 250 μm were also performed and the rapid grain coarsening observed was consistent with surface diffusion kinetics calculations. However, a rapid surface diffusion mechanism should control the neck growth at all particle sizes of interest and cannot account for the better densification of fine powders. This inconsistency, was removed by Coblenz considering that the silica layer, at the particle surfaces, might inhibit surface diffusion i.e. grain coarsening with respect to densification mechanisms [5].
    This hypothesis was recently confirmed by observations and calculations of Lebrun et al. [7, 8]. Using thermogravimetric experiments the reduction kinetics of the silica layer covering particle

surfaces was investigated and modeled. For samples heated under He-4 mol.% H_2 (2 l h^{-1}) oxidation kinetics could be sketched as follow [7]:

- For temperatures lower than 1020 °C, samples experienced a mass gain, passive oxidation of silicon occurred with the formation of solid silica, $SiO_{2(s)}$, according to reaction R_1 (equilibrium constant K_1).

$$R_1: \quad Si_{(s)} + 2H_2O_{(g)} = SiO_{2(s)} + 2H_{2(g)} \quad\quad\quad (K_1)$$

- At higher temperatures, samples started losing mass, active oxidation occurred with the formation of gaseous silicon monoxide, $SiO_{(g)}$, according to reaction R_2 (equilibrium constant K_2).

$$R_2: \quad Si_{(s)} + H_2O_{(g)} = SiO_{(g)} + H_{2(g)} \quad\quad\quad (K_2)$$

- Simultaneously with active oxidation, the silica layer started to dissociate according to reaction R_3 (equilibrium constant K_3), provided that the silica layer was thin or porous enough to allow the diffusion of silicon monoxide.

$$R_3: \quad Si_{(s)} + SiO_{2(s)} = 2SiO_{(g)} \quad\quad\quad (K_3)$$

A silica reduction front associated with grain coarsening propagated from the edge of the samples and kinetics was controlled by the diffusion of silicon monoxide released into the porous compact. Assuming a steady state diffusion of oxidizing species, the temperature at which the samples started to experience a mass loss (1020 °C) could be related to the water vapor pressure (1.8 Pa) as in the pioneer work of Wagner [9]. From the center of the compact to the reaction front, silica prevented surface diffusion and slight densification could then occur. This conclusion was confirmed by SEM observations and neck growth kinetics considerations. However, silica was reduced too rapidly and densities higher than 65 % could not be reached [8].

In this paper, the influence of the sample size on silica reduction kinetics and compact densification is investigated. With increasing the compact diameter, silica reduction and grain coarsening is prevented in the compact core and a significant increase in the final density is measured.

EXPERIMENTAL PROCEDURE

The powder consists of spherical particles of 220 nm (11.7 m^2 g^{-1}) estimated from BET specific surface area measurements (Micromeritics ASAP 2020). The morphology of the powder was observed using a Scanning Electron Microscope (SEM; LEO, STEREOSCAN 440, Cambridge, U.K.). The oxygen content (0.61 wt.%) was measured with an Instrumental Gas Analyzer (IGA; ON900, ELTRA GmbH, Neuss, Germany). The thickness of the native oxide layer was estimated from the oxygen content and the specific surface area as 0.43 ± 0.10 nm. The global amount of metallic impurities is less than 1 ppm, from Glow Discharge Mass Spectroscopy (GDMS; VG 9000, Thermo Fisher Scientific, Waltham, MA).

Cylindrical compacts (54 % relative density) were obtained by uniaxial pressing at 50 MPa followed by cold isostatic pressing at 450 MPa. Two samples of approximately 8 mm height, $\Phi_{7.5}$ and Φ_{15}, were prepared with a respective diameter of 7.5 mm and 15 mm.

Thermogravimetric experiments were performed in a SETERAM Setsys apparatus with a heating rate of 1.25 °C min^{-1} and a dwelling time of 3 h at 1350 °C. Compacts were hung up to a tungsten suspension in order to limit interactions with silicon. The furnace tube temperature was monitored with a tungsten-rhenium thermocouple and was homogenous over a range of 30 mm. He-4 mol.% H_2 commercial gas mixture was used with a typical water content of a few ppm.

After sintering sample Archimedes densities were measured and cross-sections were observed by optical microscopy and SEM.

SINTERING AND OXIDATION KINETICS OF SILICON

Sintering kinetics

The presence of silica at the particle surface strongly influences sintering kinetics. In a previous work [8], neck growth rates for two particles of radius a connected with a neck of radius x, were estimated using the formalism introduced by Ashby [10]. The main results are summarized below for the two distinct cases (Figure 1):
-   Pure silicon

A non-densifying mechanism, surface diffusion, controls sintering kinetics at all temperatures.
-   Silicon particles covered with a silica layer of thickness ε_{SiO_2}

Surface diffusion is strongly impeded. As the silicon surface diffusion coefficient, D_S [5] is replaced by the surface diffusion coefficient of silicon on silica, $D_{S\,(Si-SiO_2)}$ [11, 12], the neck growth rate is divided by $\sim 10^{16}$. Lattice diffusion (densifying) and vapor transport (non-densifying) dominate sintering kinetics in the presence of silica, depending on the temperature, the neck to particle radius ratio, $\dfrac{x}{a}$, as well as the particle size, a.

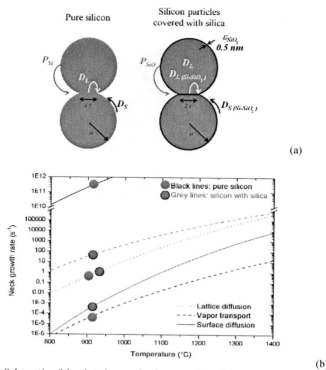

(a)

(b)

Figure 1: (a) Schematic of the sintering mechanisms considered for pure silicon and silicon particles covered with silica, D_L is the lattice diffusion coefficient of silicon into silicon, $D_{L(Si-SiO2)}$ the lattice diffusion coefficient of silicon into silica, D_S the surface diffusion coefficient of silicon onto silicon, D_S (Si-SiO2) the surface diffusion coefficient of silicon onto silica, P_{Si} the silicon vapor pressure, and P_{SiO} the silicon monoxide vapor pressure. (b) (From previous work [8]) Neck growth rate calculated for 200 nm particles at a neck to particle radius ratio of 1 %. Lattice diffusion kinetics curves are superimposed for particles with and without silica.

Silicon oxidation kinetics modeling

According to our previous works [7, 8], oxidation of silicon is controlled by reaction R_3 and diffusion fluxes of $H_2O_{(g)}$ and $SiO_{(g)}$ govern mass transport kinetics in tow distinct areas, as sketched on Figure 2:

- Outside the compact

The diffusion flux j_j^{out} of molecules j over a height z_f along the furnace tube was derived taking into account the diffusion coefficient outside the compact D_j^{out}. D_j^{out} is simply equal to the molecular diffusion coefficient D_j^{mol} estimated as a function of molecular characteristics and temperature from the semi-empiric approach of Chapman-Enskog [13, 14]:

$$D_j^{out} = D_j^{mol} \qquad (1)$$

- Inside the compact

A reduction front at the radial position r_r where reaction R_3 takes place moves from the edge ($r_r = r_c$) to the center ($r_r = 0$) of the cylindrical compact. The front rate propagation is controlled by the diffusion fluxes inside the compact j_j^{in} which are related to the diffusion coefficients inside the compact D_j^{in}. D_j^{in} is estimated as a function of the molecular diffusion coefficient D_j^{mol} and the Knudsen diffusion coefficient $D_j^{Knudsen}$ [13]:

$$D_j^{in} = \frac{p}{\tau}\left(\frac{1}{D_j^{mol}} + \frac{1}{D_j^{Knudsen}}\right)^{-1} \qquad (2)$$

Where p is the fraction of pores and τ the tortuosity.

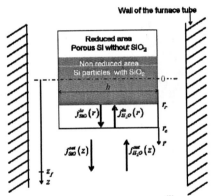

Figure 2: Schematic representation of the model reported from [8]. The gas fluxes involved in thermal oxidation of silicon compacts are represented. In the non reduced area silicon particles are covered with silica. In the reduced area silicon particles are free from oxide and gaseous molecules diffuse through the porosity. Out of the compact gaseous molecules diffuse in the furnace.

The mass loss rate is given in Eq.3 considering that H_2O arrival is responsible for a mass gain and SiO departure for a mass loss. M_j is the molar mass of a specie j, $P_{H_2O}^{z_f}$ is the water vapor pressure surrounding the sample at the z_f position and $P_{SiO}^{K_3}$ is the silicon monoxide equilibrium pressure of reaction R_3 at the reaction front position (r_r). S^{out} and S^{in} are respectively twice the furnace tube section and the lateral cylindrical sample surface. R is the gas constant, t the time and T the temperature. The reaction front position (r_r) is assessed iteratively from the weight loss, assuming that the silica is initially uniformly distributed at the particle surface throughout the sample.

$$\frac{\Delta m}{\Delta t} = \frac{S^{out}}{RT \times z_f}\left[\begin{array}{c} P_{H_2O}^{z_f} D_{H_2O}^{out} M_O - P_{SiO}^{K_3}\dfrac{D_{SiO}^{out}D_{SiO}^{in}}{D_{SiO}^{in}+D_{SiO}^{out}\ln\left(\dfrac{r_c}{r_r}\right)\dfrac{r_c}{z_f}\dfrac{S^{out}}{S^{in}}}M_{SiO} \\[3em] -P_{H_2O}^{z_f}\dfrac{D_{SiO}^{out}D_{H_2O}^{out}\ln\left(\dfrac{r_c}{r_r}\right)\dfrac{r_c}{z_f}\dfrac{S^{out}}{S^{in}}}{D_{SiO}^{in}+D_{SiO}^{out}\ln\left(\dfrac{r_c}{r_r}\right)\dfrac{r_c}{z_f}\dfrac{S^{out}}{S^{in}}}M_{SiO}\end{array}\right] \quad (3)$$

EXPERIMENTAL AND MODEL RESULTS

Mass loss curves

Experimental and expected mass losses are given in Figure 3. Expected mass losses are derived from Eq. 3, neglecting the shrinkage of the cylinder compact. The pore fraction (0.4) value is taken between the initial (0.47) and final (0.36) experimental values of the $\Phi_{7.5}$ sample. Between r_c and r_r, the pore size is fixed as 5 μm, as observed on micrographs in the reduced area [8].

The mass variation is depicted as follow:

- First, passive oxidation through reaction R_1 occurs, silica grows at the sample surface, a mass gain is measured and r_r is located at r_c.

- Once $P_{SiO}^{K_3}(T)$ equals $\dfrac{D_{H_2O}^{out}}{D_{SiO}^{out}}P_{H_2O}^{z_f}$, the silica starts to dissociate according to R_3 and passive to active transition occurs.

- At higher temperatures the mass loss rate is negative, the silica reduces until the reaction front r_r reaches the center of the compact.

  o For the $\Phi_{7.5}$ sample, the rate of mass loss suddenly collapses just before the dwell at 1350 °C. This corresponds to the end of the silica reduction since the mass loss rate becomes constant and is deduced by putting $r_r = 0$ in Eq. 3:

$$\frac{\Delta m}{\Delta t} = -\frac{S^{out}}{RT \times z_f}P_{H_2O}^{z_f}D_{H_2O}^{out}M_{Si} \quad (4)$$

  o For the Φ_{15} sample, the global mass loss is higher since the sample mass and thus the amount of silica to be reduced is also higher. A sudden collapse in the mass loss rate associated with a complete silica reduction is also predicted by the model. Experimentally, this collapse is not observed since the silica reduction is incomplete.

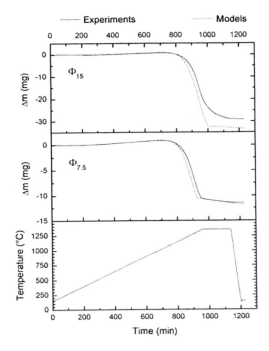

Figure 3: Experimental curves (full line) and modeled curves (dashed line) of the mass loss for the Φ_{15} and $\Phi_{7.5}$ samples associated with the thermal cycle (1.25 °C min^{-1}, 1350 °C – 3 h). Constant model parameters are $p = 0.4$, $\tau = \pi$, $\varepsilon_{SiO_2} = 0.5$ nm, $P_{H_2O}^{zf} = 1.8$ Pa, and $z_f = 40$ mm.

Sample observations

Polished cross section microstructures of the samples are given in Figure 4.
- $\Phi_{7.5}$ sample microstructure is homogeneous over the compact with large grains (2-5 μm) surrounded by coarser pores (5-10 μm). The final density of the compact is 65 % of the theoretical density.
- Φ_{15} sample microstructure is non-uniform as observed by Shaw and Heuer [2] and Möller and Welsh [3]. A relatively dense inner core (80-85 % density from image analysis) with a fine grain size (~200 nm) is surrounded by a region of coarse grain size and porosity (~65 % density from image analysis). The global density of the compact is 76 % of the theoretical density.

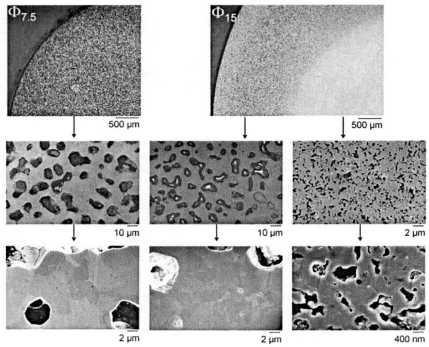

Figure 4: Optical and SEM microstructures of $\Phi_{7.5}$ and Φ_{15} polished samples.

DISCUSSION

In both compacts, surface diffusion occurring in the reduced area account for the coarse microstructure observed. During the 3 h dwell at 1350 °C, silica is still not reduced in the inner core of the Φ_{15} sample and inhibits surface diffusion and then grain coarsening. From kinetics calculations, it can be assumed that lattice diffusion is responsible for the densification observed at the center of this compact. According to Munir's approach [15], the lattice diffusion neck growth rate in the presence of a silica layer, $\dot{x}$ can be estimated using Eq. 5, where $\dot{x}_{Si}$ is the neck growth rate for pure silicon, ε_{SiO_2} the silica layer thickness and $D_{L(Si-SiO_2)}$ the lattice diffusion coefficient of silicon into silica, as measured by Brebec et al. [11] between 1110 and 1410 °C.

$$\dot{x} = \frac{D_{L(Si-SiO_2)} + \dfrac{\varepsilon_{SiO_2}}{a} D_L}{D_{L(Si-SiO_2)}\left(1 + \dfrac{\varepsilon_{SiO_2}}{a}\right)} \dot{x}_{Si} \qquad (5)$$

As shown on Figure 1, in the presence of the silica layer lattice diffusion kinetics dominates surface diffusion kinetics. However, vapor transport mechanism should also be considered. Indeed, vapor transport becomes dominant when the silicon vapor pressure is replaced by the silicon monoxide equilibrium pressure $P_{SiO}^{K_3}$ in the neck growth kinetics calculation (Figure 1). This mechanism may be responsible for the incomplete densification of the Φ_{15} inner core. In addition, the tensile stress applied by the surrounding coarse region may also prevent shrinkage of the core.

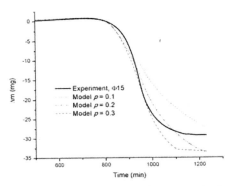

Figure 5: Effect of the porosity on the global mass loss of the Φ_{15} sample. Constant model parameters are $\tau = \pi$, $\varepsilon_{SiO_2} = 0.5$ nm, $P_{H_2O}^{z_f} = 1.8$ Pa, and $z_f = 40$ mm and a pore size of 5 μm.

The silica reduction kinetics is strongly influenced by the microstructure evolution. The porosity decrease is responsible for a diminution of the diffusion coefficient inside the compact, D_j^{in} (Eq.2) and leads to a retardation of the silica reduction as observed on the Φ_{15} experimental mass loss curve. Indeed, decreasing the pore fraction in the model leads to a better fit of the mass loss curve (Figure 5). However this fit is actually not accurate since the pore fraction is not constant inside the compact as assumed in the model derivation.

CONCLUSION

The effect of the powder compact size on silica layer reduction kinetics, microstructure evolution and final density has been studied.

Densification of pure silicon is not favored during sintering because of significant grain coarsening occurring through surface diffusion.

In small compacts, thermogravimetric experiments have shown that the silica can be rapidly reduced leading to a very porous and coarse microstructure.

In larger compacts, silica reduction is retarded, impeding coarsening and allowing lattice diffusion to occur. This leads to non-uniform final microstructures with higher densification of the core where the silica cannot be reduced.

Vapor transport is also enhanced by the presence of silica and finally prevents complete densification. Finer particles should undergo better densification by promoting lattice diffusion against vapor transport.

The silica reduction kinetics is strongly influenced by the microstructure variation. As regards silicon sintering process, the establishment of a model relating silica reduction and microstructure evolution in porous layers would be of great interest in order to control the remaining oxygen content, the porosity and eventually the electrical properties of the final material.

ACKNOWLEDGEMENTS
This work was supported by the Rhône-Alpes region through the cluster of research "Energies".

REFERENCES
[1]C. Greskovich, J. H. Rosolowski, 'Sintering of Covalent Solids', *J. Am. Ceram. Soc.*, **59** 7-8 (1976).
[2]N. J. Shaw, A. H. Heuer, 'On Particle Coarsening During Sintering of Silicon', *Acta Metall.*, **31** 55–59 (1983).
[3]H. J. Möller, G. Welsch, 'Sintering of Ultrafine Silicon Powder', *J. Am. Ceram. Soc.*, **68** 320–325 (1985).
[4]W. M. Robertson, 'Thermal Etching and Grain-Boundary Grooving of Silicon Ceramics', *J. Am. Ceram. Soc.*, **64** 9-13 (1981).
[5]W. S. Coblenz, 'The Physics and Chemistry of the Sintering of Silicon', *J. Mater. Sci.*, **25** 2754-2764 (1990).

[6]Y. Shimizu, M.Uematsu, K.M. Itoh, 'Experimental Evidence of the Vacancy-Mediated Silicon Self-Diffusion in Single-Crystalline Silicon', *Phys. Rev. Lett.*, **98** 095901 (2007).

[7]J.M. Lebrun, J.M. Missiaen, C. Pascal, 'Elucidation of mechanisms involved during silica reduction on silicon powders', *Scripta Mater.*, **64** 1102–1105 (2011).

[8]J.M. Lebrun, C. Pascal, and J.M. Missiaen, 'The role of silica layer on sintering kinetics of silicon powder compact', *J. Am. Ceram. Soc.*, Published online, DOI:10.1111/j.1551-2916.2011.05052.x, (2012).

[9]C. Wagner, 'Passivity during the Oxidation of Silicon at Elevated Temperatures', *J. Appl. Phys.*, **29** 1295–1297 (1958).

[10]M.F. Ashby, 'A first report on sintering diagrams', *Acta Metall.*, **22** 275–289 (1974).

[11]G. Brebec, R. Seguin, C. Sella, J. Bevenot, J.C. Martin, 'Diffusion du silicium dans la silice amorphe (Diffusion of silicon into amorphous silica)', *Acta Metall.*, **28** 327–333 (1980).

[12]S.H. Garofalini, S. Conover, 'Comparison between bulk and surface self-diffusion constants of Si and O in vitreous silica', *J. Non-Cryst. Solids*, **74** 171–176 (1985).

[13]R. B. Bird, W. E. Stewart, E. N. Lightfoot, Transport phenomena, John Wiley, New-York, 2001.

[14]R. A. Svehla, Technical Report TR R-132, NASA, Lewis Research Center, Cleveland, Ohio, 1962.

[15]Z. A. Munir, 'Analytical treatment of the role of surface oxide layers in the sintering of metals', *J. Mater. Sci.*, **14** 2733–2740 (1979).

Microstructural Evolution and Physical Properties

CERMETS BASED ON NEW SUBMICRON Ti(C,N) POWDER: MICROSTRUCTURAL DEVELOPMENT DURING SINTERING AND MECHANICAL PROPERTIES

A. Demoly[1], C. Veitsch[2], W. Lengauer[1], K. Rabitsch[2]

[1]Vienna University of Technology, Vienna, Austria
[2]Treibacher Industrie AG, Althofen, Austria

ABSTRACT

Ti(C,N) cermets based on a newly developed submicron Ti(C,N) grade (FSSS $\approx 0.7 \mu m$) were compared to cermets based on standard Ti(C,N) (FSSS $\approx 1 \mu m$) having the same C/N ratio. Sintering of the cermet green bodies was carried out under vacuum conditions (< 1 bar), partly under nitrogen atmosphere. In order to get insight into the development of the microstructure (i.e. phase formation and evolution during sintering, porosity), the sintering process was interrupted at specific temperatures (1175, 1300, 1400 and 1460°C, respectively) by fast cooling. The changes in the microstructure were monitored by XRD, SEM, light-optical microscopy and magnetic properties. XRD analysis shows the total dissolution of the (Ta,Nb)C and WC during heating and the formation of three fcc phases at the end of the sintering cycle. One corresponds to the binder phase, the second can be assigned to the rim phase and exhibits a complex solid solution (Ti,Ta,Nb,W)(C,N) and the third to undissolved Ti(C,N) cores. The small difference in the Ti(C,N) starting powder grain size was found to have a significant impact on the microstructure. The submicron powder shows higher activity upon sintering, therefore, a larger amount of rim phase and a further dissolution of the Ti(C,N)-rich cores are observed. The microstructural changes induced by using submicron Ti(C,N) cause higher magnetic specific saturation ($4\pi\sigma$), coercive force (HcJ) and hardness (HV10), whereas the Palmqvist fracture toughness (K_{IC}) decreases slightly.

INTRODUCTION

The microstructure of Ti(C,N)-based cermets is typically characterised by a core-rim structure. The rim is composed of an inner and an outer part, which are formed during solid-state and liquid-phase sintering, respectively. In Ti(C,N)-based cermets the cores stem from the undissolved Ti(C,N) particles, while an homogenised carbonitride phase forms the rim. The main parameter controlling the microstructure formation is the diffusion of the elements. Therefore, using smaller starting Ti(C,N) particle size may lead to a significant modification of the microstructure and thus of the properties of cermets[1].

The present study compares the development of the cermet microstructures during sintering using interruption of the sintering process and characterization by SEM, XRD as well as magnetic properties measurements. Two powders were used, the first is a conventional fine Ti(C,N) powder the second a newly developed submicron Ti(C,N) powder.

EXPERIMENTAL

The chosen material specifications belong to the material class HT-P10 (ISO513 1992-06), which contains a large amount of Ti(C,N) and has a low content of other carbides. This application class fits particularly for the machining of steel and cast steel with high cutting speed and low feed rate.

Starting materials

The cermets composition is listed in Table I, while the properties of the individual powders are listed in Table II. Two Ti(C,N) powder grades – a conventional Ti(C,N) powder (FSSS = 1μm) and a

submicron Ti(C,N) powder (FSSS = 0.7μm) – both with a [C]/([C]+[N]) ratio of 0.5 – were used to prepare cermet mixtures (Table II, Figure 2). Both Ti(C,N)-grades as well as the other carbides are industrial grades by Treibacher Industrie AG. Figure 1 shows a SEM image of the Ti(C,N) powders. The binder powders Co and Ni were supplied by Umicore and Inco, respectively.
The grain size was determined using the Fischer Sub Sieve Sizer (FSSS) (Table II) and diffraction laser particle size analyser (Figure 2). The O, C and N contents were measured by carrier gas hot extraction.

Table I. Composition of the cermets I and II

		fine Ti(C,N)	submicron Ti(C,N)	(Ta,Nb)C	WC	Cr₃C₂	Co	Ni
Grade A	I (wt%)	59	-	10	16	1	6.5	6.5
	II (wt%)	-	59	10	16	1	6.5	6.5

Table II. Properties of the individual powders

	fine Ti(C,N)	submicron Ti(C,N)	(Ta,Nb)C	WC	Cr₃C₂	Co	Ni
Carbon (wt%)	10.59	10.75	8.10	6.17	13.22	0.15	0.13
Nitrogen (wt%)	10.79	10.60	0.01	-	0.02	-	-
Oxygen (wt%)	0.59	1.08	0.24	0.19	0.53	0.55	0.05
Grain size (FSSS in μm)	1.0	0.7	1.3	0.6	1.9	0.9	1.3

Figure 1. Ti(C,N) powders, conventional (left) and submicron (right), SEM, SE, 20000x

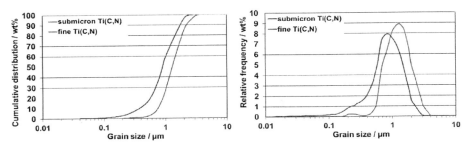

Figure 2. Ti(C,N) powders, cumulative (left) and differential (right) grain size distribution

Preparation of green bodies
    After weighing the powders, the blends were mixed and milled for 72h in a ball-mill filled with hardmetal balls and cyclohexane. The obtained cermet powders were dried and agglomerated and then pressed with 150MPa to cylindrical bodies (8mm height and 12mm in diameter) in an uniaxial press.

Dilatometry
    The shrinkage of the samples was investigated by dilatometry (DIL801, Bähr-Thermoanalyse GmbH) following the sintering profile depicted in Figure 3 but under vacuum and with slightly shorter dwell times.

Sintering
    The samples were placed in a graphite crucible and sintered in an induction furnace first under vacuum and then under nitrogen atmosphere (Figure 3). The sintering process was interrupted by fast cooling at 1175°C, 1300°C, 1400°C and 1460°C, respectively, after different dwell times (Figure 3, Table III) allowing a chronological observation of the microstructure evolution, phase formation and of the magnetic properties. After interrupting the sintering process, XRD, light-optical microscopy (LOM) and scanning electron microscopy (SEM) were applied to assess the formed phases. Furthermore, light optical microscopy (LOM) images were made with an inverted reflected light microscope (Olympus Co.), and image analysis with the program AnalySIS Ver. 5.0. The polished samples were etched using an basic oxide polishing silicate suspension to remove the binder and to obtain an higher contrast of the grain boundaries. A scanning electron microscope (FEI Quanta 200FEG) using a back scattered electron filter was employed to inspect the microstructure as well as the grain size of the cermets. The calculation of the core-to-rim ratio was made using the program ImageJ 1.43u on basis of the grey scale difference between core (dark) and rim (bright). The coercive force (HcJ) and the weight specific magnetic saturation ($4\pi\sigma$) were measured using a Koerzimat CS 1.096 (Förster). After final sintering the hardness (HV10) and fracture toughness (K_{IC}) of the samples were measured using the Shetty equation.

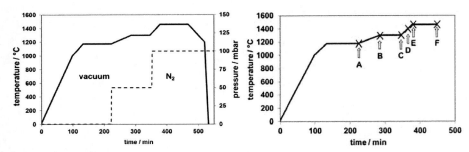

Figure 3. Sintering profile (left) and sintering interruption points (right)

Table III. Sintering interruption points

	A	B	C	D	E	F
Temperature in °C	1175	1300	1300	1400	1460	1460
Time in min	90	0	60	0	0	90

RESULTS

Dilatometry

The shrinkage curves show a higher shrinkage for cermet II with submicron Ti(C,N) from the beginning of the first dwell up to the formation of the liquid phase, during the third dwell. The solid-state sintering is obviously further advanced for cermet II with submicron Ti(C,N) than for cermet I with fine Ti(C,N). Generally, the beginning of the shrinkage correlates with the rearrangement of powder particles. The introduction of finer Ti(C,N) powder in cermet II causes faster diffusion and therefore, higher shrinkage from 1175°C up to around 1400°C (appearance of the liquid phase). The formation of the liquid phase at 1400° marks the end of the solid-state sintering regime and the balancing of the shrinkage for both cermets.

The total shrinkage (Table IV) as well as the shrinkage rate (Figure 4) are identical for both cermets and therefore, independent of the grain size of the used Ti(C,N) powder. In addition, the temperature of the shrinkage rate maximum is almost identical for both cermet grades (Table IV).

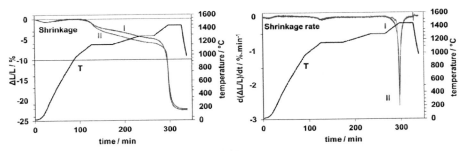

Figure 4. Shrinkage (left) and shrinkage rate (right) of cermets I and II

Table IV. Shrinkage rate and overall shrinkage of cermets I (with fine Ti(C,N) and II (with submicron Ti(C,N))

Cermet	Main shrinkage rate maximum °C	Total shrinkage %
I	1425	22.3
II	1426	22.6

Apparent porosity - light optical microscopy (LOM)

The evolution of the microstructure during the sintering process of stages C to F is shown in Figure 5 and 6 (for porosity quantification of all stages and SEM analysis see below). During heating from 1175°C to 1300°C homogenisation occurs alongside the formation of new phases. At stage C (1300°C – 60 min) no large difference between the two cermet grades can be observed. A large amount of rim phase had formed and the porosity had further declined. At 1400°C (stage D), a dramatic modification of the porosity is observed. Cermet I with fine Ti(C,N) shows fewer but larger irregular pores whereas cermet II with submicron Ti(C,N) has many and much smaller pores with rounded shape. This pore structure continues to stage E (1460°C – 0 min), however, with a significant drop in apparent porosity. Finally, the porosity disappears completely during liquid-phase sintering at 1460°C between E (0 min) and F (90 min) in both grades. In Figure 7, the porosity surface area (from image analysis) of both samples is presented for all stages. Cermet II with submicron Ti(C,N) exhibits a lower porosity than cermet I between 1175°C and 1400°C (D). Obviously, the densification of cermet II takes largely place between 1175°C (A: 33%) and 1300°C (B: 9%), whereas for cermet I with fine Ti(C,N) the densification needs longer and higher temperatures. The porosity of cermet I after 90 min dwell time at 1300°C (C: 12%) is still slightly higher than that of cermet II at 1300°C without dwell time (B: 9%). During the dwell time at 1300°C (B-C) the porosity of sample II decreases slightly from 9 to 7%, while the porosity of cermet I decreases dramatically from 27 to 12%. Above 1400°C (D), the porosity of both cermets is nearly the same and full densification is achieved at 1460°C after 90 min (F).

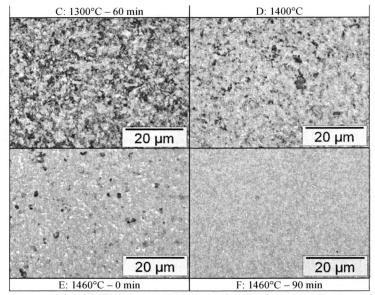

Figure 5. Microstructure evolution during sintering, cermet I

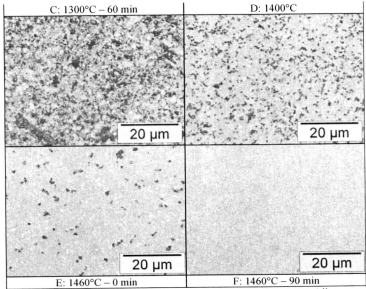

Figure 6. Microstructure evolution during sintering, cermet II

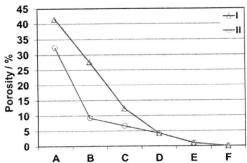

Figure 7. Porosity upon sintering for cermet I (with fine Ti(C,N)) and II (with submicron Ti(C,N))

Scanning electron microscopy

As Figure 8 and 9 show the evolution of the microstructure using a larger magnification of the SEM, several new features can be detected. At 1175°C (A) some binder pools are present. The homogenisation of the binder is first completed above 1400°C (D-F), which correlates with an increase of the shrinkage rate and the subsequent formation of the liquid phase (Figure 4).

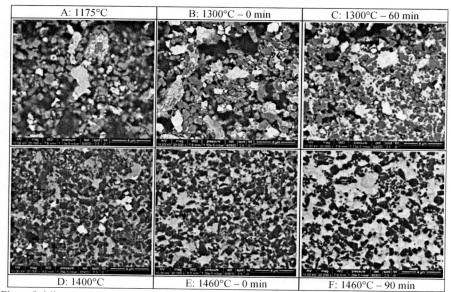

Figure 8. Microstructure evolution of cermet I (fine Ti(C,N)), SEM, BSE, 20000x

The Ti(C,N) particles (their comparatively low atomic number causes the particles to appear as almost black in back-scattered SEM-images) are present during the entire sintering cycle, although some dissolution takes place (Figure 8, 9). At stages where Ti(C,N) can be unambiguously identified (D to F) a much finer particle size is visible in cermet II. Image analysis reveals that the Ti(C,N) fraction decreases faster with submicron Ti(C,N) as for fine Ti(C,N) (Figure 10). This shows that the reactivity of submicron Ti(C,N) powder is larger than that of fine Ti(C,N) powder[2]. The WC fraction (the brightest particles in the back-scattered SEM-images) declines during sintering, while a homogenised outer rim phase composed of all elements except the binder phase (intermediate grey) is formed (Figure 8, 9 and 10). The hexagonal WC is dissolved in both cermet grades between stage D and E. Some inverse grains with bright core (W-rich) and darker rim (Ti-rich) are occasionally formed during the dwell time at 1460°C (F).

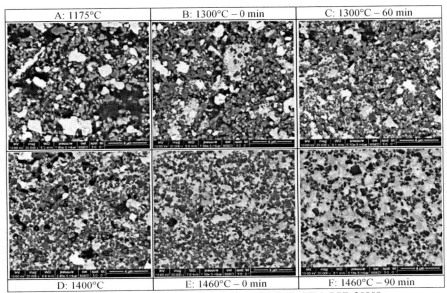

Figure 9. Microstructure evolution of cermet II (submicron Ti(C,N), SEM, BSE, 20000x

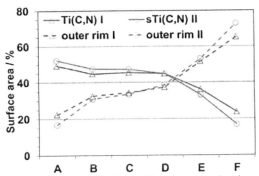

Figure 10. Surface area of Ti(C,N) core- (continuous line) and outer rim-phase (broken line) during sintering for cermet I (triangle) and II (circle)

Phase formation investigated by XRD

Figure 11 shows the XRD patterns of both cermets at the various stages of sintering (A-F). It is apparent, that for both cermets the (200) line of (Ta,Nb)C at 40.5° can only be identified at stage A

(1175°C). At the next stage B (1300°C – 0 min) no more (Ta,Nb)C is present. Between 1175°C and 1300°C this phase homogenised with parts of Ti(C,N) to form a homogeneous fcc outer rim phase. At the formation temperature range between 1175 and 1300°C both cermets exhibit a large porosity, therefore their nitrogen activity is low[3] and the so-formed rim contains primarily a large amount of the metals Ti, Ta, Nb together with C. This multi-component rim phase has an increased lattice parameter as compared to Ti(C,N) (Table V). With increasing temperature a further dissolution of Ti(C,N) and WC leads to a decrease of the lattice parameter of the homogenised phase after stage D (Figure 12, center). The intake of W,Ti and N, makes the lattice smaller and shifts the corresponding XRD-lines to larger diffraction angles (Figure 11). With increasing temperature and mainly in presence of liquid phase above 1400°C (D-F) the homogenisation increases, leading finally to an overlap of the diffraction lines of the homogenised outer rim phase with that of the remaining Ti(C,N) core.

Obviously, the lattice parameter of Ti(C,N) in cermet I remains constant during sintering, while in cermet II it increases slightly up to 1400°C. Supposing that Ta, Nb and W do not diffuse into the Ti(C,N) particles, it implies an increase of the [C]/[N] ratio, firstly by diffusion of dissolved C from the binder phase into Ti(C,N)[4] and secondly by outgassing of N_2[5]. During liquid-phase sintering, the lattice parameter of Ti(C,N) decreases for both grades because of the higher N-activity, leading to a lower [C]/[N] ratio. The higher reactivity of submicron Ti(C,N) leads to a larger decrease of the [C]/[N] ratio and therefore to a lower lattice parameter (Table IV, Figure 12, left). The lattice parameter of the binder phase changes dramatically from room temperature to 1175°C (Figure 12, right), after that it remains practically constant. The increase is primarily due to the dissolution of WC[6] and (Ta,Nb)C[7,8]. Afterwards, the lattice parameter of the binder for both cermets is nearly equal - slight variations are caused by different amounts of dissolved elements.

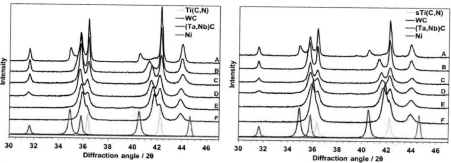

Figure 11. XRD patterns of cermet I (left) and II (right) after interrupted sintering compared with starting powders

Table V. Lattice parameters of Ti(C,N), binder and homogenised outer rim phase during sintering
*blend: binder phase from the lattice parameters of Co and Ni using Vegard's law

	Lattice parameter / Å					
	Core - Ti(C,N)		Outer rim phase		Binder phase	
	I	II	I	II	I	II
Blend	4.284	4.286	-	-	3.534*	3.534*
A: 1175°C	4.284	4.286	4.355	4.351	3.572	3.574
B: 1300°C – 0 min	4.284	4.285	4.378	4.372	3.573	3.572
C: 1300°C – 60 min	4.285	4.287	4.359	4.356	3.571	3.571
D: 1400°C	4.285	4.289	4.353	4.352	3.574	3.575
E: 1460°C – 0 min	4.284	4.287	4.341	4.343	3.577	3.575
F: 1460°C – 90 min	4.282	4.280	4.323	4.320	3.573	3.572

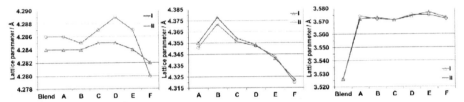

Figure 12. Lattice parameter of Ti(C,N) (left), the outer rim phase (center) and the binder phase (right)

Magnetic and mechanical properties

When interrupting the sintering process, it is assumed that the binder is not given enough time to change its composition due to the high cooling speed and that the magnetic properties remain constant.

After a significant increase of the magnetic saturation ($4\pi\sigma$) from stage A (1175°C – 90 min) to stage B (1300°C – 0 min), due to the formation of a binder phase skeleton, the magnetic saturation remains approximately constant (stage B to E) (Table VI, Figure 13). The slightly higher magnetic saturation of cermet II in stages C (1300°C – 60 min) to F (1460°C – 90 min) might originate in the lower amount of dissolved elements in the binder. The hard phase components show higher reactivity among themselves and therefore, a slightly lower dissolution in the binder than in cermet I[9]. This is in correspondence with the advanced formation of outer rim phase in cermet II, as compared to that of cermet I (Figure 10), which is especially true for the last step F (1460°C – 90 min). The increase in magnetic saturation at 1460°C from E (0 min) to F (90 min) is due to the redissolution of alloying elements, present in the binder phase, back into fcc-hard phase particles, aided by the presence of liquid phase.

The coercive force (HcJ) increases significantly from A (1175°C – 90 min) to D (1400°C), where the liquid phase is formed (Table VI, Figure 13). Then, it remains relatively stable at stages D (1400°C) to F (1460°C – 90 min). This stability of HcJ at high temperature is an indication of the stability of the hard phase and the absence of grain growth. Cermet II has a higher HcJ than cermet I due to the finer grain size of hard particles in cermet II.

The hardness (HV10) of cermet I and II after sintering is 1820 and 1850, while their fracture toughness (K_{IC}) is 8.0 and 7.5, respectively. Cermet II has a higher hardness since more fine Ti(C,N) particles remain in the final microstructure (Hall-Petch effect). Furthermore there is a larger amount of outer rim phase (Ti,W,Ta,Nb)(C,N) (25-26 GPa HV0.1)[10] present in Cermet II, which is harder than Ti($C_{0.5}N_{0.5}$) (23 – 24 GPa HV0.1)[11] and which therefore increases the total hardness.

Table VI. Magnetic properties evolution during sintering

	$4\pi\sigma$ / $\mu T \cdot cm^3 \cdot kg^{-1}$		HcJ / $kA \cdot m^{-1}$	
	I	II	I	II
A: 1175°C	57.8	60.4	4.76	5.32
B: 1300°C - 0 min	98.8	96.3	12.37	14.56
C: 1300°C - 60 min	100.9	104.4	14.37	17.8
D: 1400°C	101.6	103.6	20.07	23.29
E: 1460°C - 0 min	100.0	104.9	18.48	23.43
F: 1460°C - 90 min	107.6	115.4	18.85	23.09

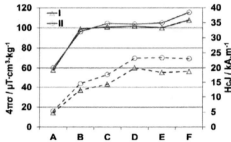

Figure 13. Magnetic saturation $4\pi\sigma$ (continuous line) and coercive force (broken line) of cermets I (triangle) and II (circle) during sintering

CONCLUSIONS

The current study highlights the effects of submicron Ti(C,N) powder in comparison with fine Ti(C,N) powder in the sintering process of Ti(C,N)-rich cermet alloys. Dilatometry experiments reveal a higher shrinkage during solid-state sintering for the cermet-grade containing submicron Ti(C,N) powder, because of the powders higher reactivity, while the total shrinkage after liquid-phase sintering remains unchanged. By interrupting the sintering cycles at various temperatures it was found that the more reactive submicron Ti(C,N) powder causes further dissolution of Ti(C,N) cores during liquid-phase sintering. Thus, a lower fraction of Ti(C,N) and a larger amount of a homogeneous outer rim phase, composed of dissolved elements, remain after sintering. A comparison of the cermets' magnetic properties confirm a smaller grain size of cermets made of submicron Ti(C,N) through all stages of sintering and a higher fraction of the homogeneous outer rim phase. This leads to a higher hardness of the cermet made from submicron Ti(C,N).

REFERENCES

[1]A. Demoly, W. Lengauer, C. Veitsch, K. Rabitsch, *Effect of submicron Ti(C,N) on the microstructure and the mechanical properties of Ti(C,N)-based cermets*, Int. J. Refract. Met. Hard Mater., **29**, 716-723 (2011).

[2]J. Jung, S. Kang, *Effect of ultra-fine powders on the microstructure of Ti(CN)-xWC-Ni cermets*, Acta Mater., **52**, 1379-1386 (2004).

[3]P. Lindahl, P. Gustafson, U. Rolander, L. Stals, H.-O. Andrén, *Microstructure of model cermets with high Mo or W content*, Int. J. Refract. Met. Hard Mater., **17**, 411-442 (1999).

[4]P. Ettmayer, H. Kolaska, W.Lengauer, K. Dreyer, *Ti(C,N) cermets – metallurgy and properties*, Int. J. Refract. Met. Hard Mater., **13**, 343-351 (1995).

[5]L. Chen, W. Lengauer, P. Ettmayer, K. Dreyer, H.W. Daub, D. Kassel, *Fundamentals of liquid phase sintering for modern cermets and functionally graded cemented carbonitrides (FGCC)*, Int. J. Refract. Met. Hard Mater., **18**, 307-322 (2000).

[6]J. Zackrisson, H.-O. Andrén, *Effect of carbon content on the microstructure and mechanical properties of (Ti,W,Ta,Mo)(C,N)–(Co,Ni) cermets*, Int. J. Refract. Met. Hard Mater., **17**, 265-273 (1999).

[7]H. Yoshimura, T. Sugizawa, K. Nishigaki, H. Doi, *Reaction occurring during sintering and the characteristics of TiC-20TiN-15WC-10TaC-9Mo-5.5Ni-11Co Cermet*, Int. J. Refract. Met. Hard Mater., **2**, 170-174 (1983).

[8]M. Qian, L.C. Lin, *On the disappearance of Mo2C during low temperature sintering of Ti(C,N)-Mo2C-Ni cermets*, J. Mater. SCi. 34, 3677-3684 (1999).

[9]D.-G. Ahn, K.-W. Lee, J.-W. Lee, M. Sharon, *Study on magnetic-mechanical properties of Ti(C,N)-based cermet with sintering conditions*, Proc. WorldPM2004, Vienna (A), 537-542, EPMA, Shrewsbury, UK (2004).

[10]R. Königshofer, A. Liersch, W. Lengauer, T. Koch, M. Scheerer, W. Hohenauer, *Solid-state properties of binary, ternary and quaternary transition metal carbonitrides*, World PM2004, PM Tool Materials, Vienna, Vol. 3, 593-598 (2004).

[11]W. Lengauer, S. Binder, K. Aigner, P. Ettmayer, A. Guillou, J. Debuigne, G. Groboth, *Solid state properties of group IVb carbonitrides*, J. Alloys Comp. **217**, 137-147 (1995).

GRAIN GROWTH OF β-Si$_3$N$_4$ USING Y$_2$O$_3$ AND Al$_2$O$_3$ AS SINTERING AIDS

Leonel Ceja-Cárdenas[1,2], José Lemus-Ruiz[1] Sebastián Díaz de la Torre[2], Egberto Bedolla-Becerril[1]

[1] Instituto de Investigaciones Metalúrgicas, UMSNH., Edif. "U", CU., Apdo. Postal 888, C.P. 58000, Morelia, Mich., México.
[2] Instituto Politécnico Nacional, CIITEC. Cerrada Cecati s/n Col. Sta. Catarina, C.P. 02250 Azc., D.F., México.

ABSTRACT

The mechanical properties of silicon nitride ceramics are closely linked to the α-β phase transformation. Spark plasma sintering (SPS) technique has been used to densify pure α-Si$_3$N$_4$ commercial powder, using Y$_2$O$_3$ and Al$_2$O$_3$ as additives; from 2.5 and 5.0 wt% and 1.5 and 3 wt%, respectively. Such powder admixtures were previously spray-dried at 160°C in such a way that powder was thoroughly homogenized. The sintering treatment included: 3 to 20 min holding time, 38 MPa axial load, sintering temperature of 1500°C and heating rate of 300°C/min. The maximum relative density developed on studied specimens is nearly 99% of theoretical value, which corresponds to 2M-series specimens and could only be attained once the β-phase nucleated from the α-silicon nitride matrix. X-ray diffraction (XRD) and scanning electron microscope (SEM) analyses confirmed the presence of two mains phases in the resultant microstructures: α- and β-Si$_3$N$_4$. Both the β-Si$_3$N$_4$ grain growth and the role that different amount of additives can play during SPS-sintering are analyzed.

INTRODUCTION

Spark Plasma Sintering (SPS) technique is a non-conventional powder consolidation method that enables ceramics to be fully densified at relatively low temperatures at very short times. This behavior is associated to the spark discharge effect that leads to local heating between powder particles, which in turn might promote the material transfer and/or atomic diffusion (depends on the nature of the material being sintered) via chemical activation of the particles surface chemical potential, resulting in higher densification. One of the major advantages of this process is the possibility of setting up rapid heating and cooling rates. This fact opens an opportunity to closely follow the materials phases' transformation that is otherwise very difficult to follow by conventional sintering techniques [1-2].

Silicon nitride (Si$_3$N$_4$) is a chemical compound now known to undergo three crystalline forms, namely: α (trigonal), β (hexagonal) and γ (cubic) [3]. To achieve the α- to β-phase transformation in Si$_3$N$_4$-base ceramics it is necessary that the sintering aids used (metallic oxides, M$_x$O$_y$) react with the SiO$_2$ traces typically present on the surface of the α-Si$_3$N$_4$ powders to form an eutectic liquid, so that small α particles dissolve in this liquid and can precipitate as β-Si$_3$N$_4$. In this case, the excessive liquid phase developed in sintered ceramic can be perceived as remaining liquid, which is kinetically frozen and conventionally called intergranular glassy film IGF, due to its location in the microstructure. The later taking place according to the reaction (1):

$$\alpha\text{-Si}_3\text{N}_4 + \text{SiO}_2 + \text{M}_x\text{O}_y \rightarrow \beta\text{-Si}_3\text{N}_4 + \text{M-Si-O-N(IGF)} \tag{1}$$

The resulting microstructure of these ceramics after the sintering process is similar to that of whisker-reinforced composites, exhibiting anisotropic growth of rod-like β-Si$_3$N$_4$ grains, which act as interlocking elements throughout the glassy phase. This microstructural development in the Si$_3$N$_4$-ceramics is one of the most important factors associated to their excellent mechanical properties, exhibited either at ambient or elevated temperatures [4-6].

The sintering process of silicon nitride is a phenomenon well known and numerous microstructural observations have shown that the β-Si₃N₄ grains grow as elongated hexagonal prisms. For years several investigations have been devoted to controlling the size, shape, morphology, and distribution of the β-grains in Si₃N₄-ceramics, in order to improve the mechanical properties [7-9]. Using the gas-pressure-sintering technique and 6 wt% Y₂O₃, 2 wt% Al₂O₃, and 0-7 wt% β-Si₃N₄ seeds as sintering additives, C. Liu [10] reported sintering of α-Si₃N₄ ceramics. He found that the number of enlarged grains increases consistently with a small amount of β-Si₃N₄ seeds. M. Belmonte et al [11] used the hot-pressing technique to evaluate the Si₃N₄ grain growth effect when the Al₂O₃ content was increased from 2 to 4 %wt, using 5 wt% Y₂O₃ and 0-4.7 wt% of elongated β-seeds. They reported that by increasing the Al₂O₃ content the grain growth in non-seeded samples significantly enhanced, whereas the opposite occurred in seeded materials. They concluded that the grain growth is influenced by additive composition especially through the liquid viscosity at the sintering temperature. Therefore, there are some experimental factors influencing the microstructure's behavior, which may cause the so-called exaggerated or abnormal grain growth of β-Si₃N₄. Others factors that are also believed to contribute to abnormal grain growth, include: inhomogeneous distribution of liquid phases [12], β-Si₃N₄ grain morphology [13], the β-Si₃N₄ nuclei density [14], and the β-Si₃N₄ grain size distribution within the starting powder [15]. Although, the major cause of the growth in those ceramics is not yet known, it is generally accepted that growth of elongated grains is predominantly diffusion-controlled. The factors governing such grain growth are still a matter of discussion.

On the other hand, it is well known that the main role of the eutectic liquid phase present during the sintering process which enhances densification of Si₃N₄-ceramics through atoms diffusion within the liquid phase and subsequently acts as catalyzer of the α- to β-phase transformation, reduces the specimen's porosity. The usage of sintering additives is thus a common practice in silicon nitride ceramics to facilitate densification via glassy phase formation [16]. Controlling the amount of additives however is important since the final properties of resulting ceramics, such as thermal conductivity, fracture toughness, and others parameters can substantially be affected. In the present work, the amount effect of additives used to develop the β-grain growth in Si₃N₄-ceramics, sintered by the Spark Plasma Sintering technique is reported.

EXPERIMENTAL PROCEDURE

The powder mixtures set in this work were prepared from commercial high purity alpha silicon nitride powder (99.9%), Toshiba Ceramics Co., Ltd. USA, of 1.1μm average particle size. Oxide additives such as Y₂O₃ (99.9%) Molycorp Minerals, USA and Al₂O₃ (99.99%) Taimei Chemicals Co., Ltd., Tokio, Japan were used as sintering aids. To analyze the amount effect of additives on the grain growth upon sintering, two compositions were prepared (see Table 1). Selected compositions were first mixed using Spray Drying technique (Mini-Spray Dryer ADL 31) using flow velocity of 0.4 lit/h while temperature inside the drying chamber was set at 160°C. Powder mixtures were pre-compacted using graphite die-matrix of inner diameter of 10 mm and sintered by a spark plasma sintering apparatus (Model: SPS-1050) at 1500°C for 3-20 min holding time. The sintering temperature was measured using a pyrometer focused on the surface of the graphite die. Uniaxial load pressure of 38 MPa and vacuum atmosphere of 0.5 Pa were applied from the start to end of the sintering cycle.

Table I. Code and chemical composition of SPS-treated specimens (wt%).

Mixture	α-Si₃N₄	Y₂O₃	Al₂O₃
M1	96	2.5	1.5
M2	92	5	3

The bulk density of sintered bodies was measured by the Archimedes's principle using distilled water. X-ray diffraction (XRD) analysis was used for microstructural phase identification. The morphology and physical condition of fractured surfaces of studied materials were observed using JEOL JSM- 6400 scanning electron microscope (SEM). Specimens were fractured and chemically etched with hydrofluoric acid at room temperature for 30 min to reveal microstructural details.

RESULTS AND DISCUSSION

Density measurements revealed high compaction level attained on specimens after sintering. The maximum relative density attained in the M1-series specimens sintered at 1500°C for 20 min was about 90% whereas 99% for the M2-series treated at 1500°C for 3 and 20 min. Which confirmed that at this stage of the sintering process, the more amount of additives promotes larger densification level developed on those specimens. G. Ling et al [17] reported the relative density of pressureless-sintered Si_3N_4-5 wt% MgO-Y_2O_3 ceramics as a function of the Y_2O_3 amount. Their Si_3N_4 ceramic bodies attained better densification when the Y_2O_3 amount was ranged from 0 to 4 wt% and increased porosity from 5 to 6 wt% Y_2O_3. They concluded that only moderate amount of liquid phase during sintering process leads to high density in Si_3N_4-ceramics, but too much or less liquid phase lead to poor density.

Fig. 1 shows X-ray diffraction patterns of the M1-series specimens sintered at 1500°C for 5 and 20 min, respectively. The 5 min sample reveals that β-phase transformation is not yet achieved on the precursor powder (Fig. 1a). However, the presence of both phases in Si_3N_4-ceramics becomes evident, as detected after 20 min holding time (Fig. 1b). These results suggest that the M1-series specimens, having 4 wt% total additives (Y_2O_3-Al_2O_3) cannot undergo complete α- to β-phase transformation, independently of sinter holding-time. Possibility, due to insufficient eutectic liquid amount formed.

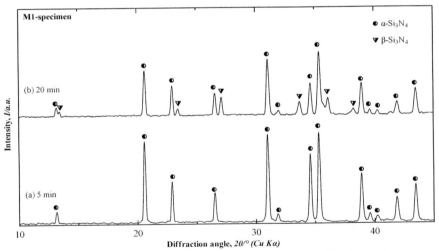

Figure 1. XRD-patterns of M1-specimens sintered at 1500°C with holding sintering times of 5 and 20 min, respectively.

Fig. 2 shows the X-ray diffraction patterns of M2-series specimens sintered for 3 and 20 min at 1500°C. In the first case, the α- to β-phase transformation initiated faster than M1-series. However, the complete transformation cannot be achieved in spite of the fact that sintering time lasted for 20 min (Fig. 2b). A comparison made between X-ray diffraction patterns of Fig. 1(a) and Fig. 2(a) reveals significant differences. Whereas the M1-series specimen, having 4 wt% total additives presented no phase transformation at 1500°C for 5 min, specimens of M2-series having twice the amount of additives revealed already β-phase diffraction peaks. Evidently, the additives amount as set in these experiments confirms previous research data on the influence of phase transformation on α- to β-phase. Fig. 2(b) also disclosed unidentified diffraction peaks from 28 – 31 (2θ) degrees and others at around 32 and 37 (2θ) degrees, which need further indexation.

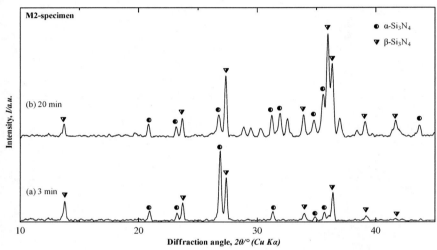

Figure 2. XRD-patterns of M2-specimens sintered at 1500°C with holding sintering times of 3 and 20 min, respectively.

Fig. 3 shows SEM images of the fractured surface of Si₃N₄ samples having different sintering conditions. Fig. 3 (a-b) corresponds to the M1-series specimens sintered at 1500°C for 20 min. Fig. 3 (c-d) shows the M2-series specimens sintered at 1500°C for 3 min. A comparison of these SEM pictures reveals important morphological differences between specimens sintered. Fig. 3 (a) shows a representative view of sintered bodies with no β-grain growth on the precursor powder in spite of the fact that the sintering time lasted for 20 min. Fig. 3(b) is a close up view of fractured zone. At the center of this picture not only a small β-Si₃N₄ grain is observed but also some α-Si₃N₄ crystals on its way to adopt hexagonal forms, both of them are surround by equiaxed grains. However, the majority of grains were equiaxed and its microstructure almost the same as that of the starting powder, indicating little phase transformation. Fig. 3(c) shows a representative view of changes suffered by the α-Si₃N₄ particles along the SPS-treatment for 3 min. It is possible to observe that some particles adopted the typical rod-like shape of hexagonal β-Si₃N₄ crystals. Fig. 3(d) is a close up view of β-Si₃N₄ rod, which had the largest amount of additives during sintering process. These results suggested that specimen of the M1-series, having 4 wt% total additives presented a smaller β-grain growth than the specimen of the M2-series, which have 8 wt% total additives. Therefore, it is evident that the additive amount has significant effect on grain growth of β-Si$_3$N$_4$ for the conditions used in this work. The larger amount of additives used to densify Si$_3$N$_4$ powder (8wt% in this work) the more amount of liquid phase formed during the SPS-sintering process. The liquid phase promotes atoms migration through it supporting the material transfer and β-grain growth. Apparently, the grain growth mechanism is controlled by elements diffusion through the liquid phase. Apparently, the formation of a liquid phase is a necessary condition for obtaining dense Si$_3$N$_4$ specimens. However, an inappropriate amount of additives used during the sintering process could kinetically frozen the microstructure promoting the vitreous phase formation. This situation could not be beneficial to the properties of Si$_3$N$_4$-ceramics. On the contrary, if the additives amount set is not sufficient then the phase transformation and the relative density could not be achieved. Therefore, optimizing the amount of Y$_2$O$_3$ and Al$_2$O$_3$ additives to sinter Si$_3$N$_4$ is an experimental task to find out upon SPS-sintering.

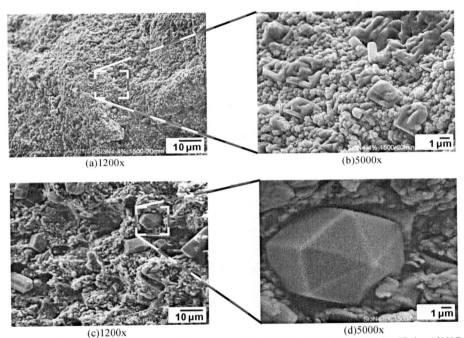

(a)1200x (b)5000x

(c)1200x (d)5000x

Figure 3. (a) and (b) Fracture surface of M1-specimens (96 wt%Si$_3$N$_4$-2.5Y$_2$O$_3$-1.5Al$_2$O$_3$) densified at 1500°C for 20 min. Irregular grains seen in (a – b) are α-Si$_3$N$_4$ grains. Figure 3. (c) and (d) Fracture surface of M2-specimens (92 wt%Si$_3$N$_4$-5Y$_2$O$_3$-3Al$_2$O$_3$) densified at 1500°C for 3 min. Hexagonal rod seen in (c – d) is β-Si$_3$N$_4$ grain.

CONCLUSIONS

The spark plasma sintering technique was used to densify pure α-Si$_3$N$_4$ commercial powder having Y$_2$O$_3$ and Al$_2$O$_3$ additions, from 2.5 to 5.0wt% and 1.5 to 3wt%, respectively. The grain growth of β-Si$_3$N$_4$ was analyzed as a function of the additives amount. On the bases of these experiments the following conclusions can be summarized:

1. Relative density measurements indicate a better sintering level reached on the M-2 series specimens. The M-2 series contained twice amount of additives than M-1 series. The more amount of additives leads to larger amount of liquid phase formed upon sintering, which apparently facilitates densification of the SPS-treated Si$_3$N$_4$-ceramics.

2. X-ray diffraction analyses revealed that the complete α- to β-phase transformation cannot be achieved in the studied specimens under the experimental conditions set in this work.

3. The results of this work confirm a clear influence of the additives amount on the α- to β-phase transformation and grain growth upon SPS-sintering.

4. SEM analyses revealed that specimens of the M1-series having 4 wt% total additives (Y$_2$O$_3$-Al$_2$O$_3$) undergo smaller β-grain growth as compared to counterpart specimens of the M2-series having 8 wt% additives.

REFERENCES

[1] Z. Shen and M. Nygren, Kinetic Aspects of Superfast Consolidation of Silicon Nitride Based Ceramics by Spark Plasma Sintering, *J. of Mater. Chem.*, **11** 204-207 (2001).

[2] T. Nishimura, X. Xu, K. Kimoto, N. Hirosaki, and H. Tanaka Fabrication of silicon nitride nanoceramics—Powder preparation and sintering: A review, *Sci. Technol. Adv. Mater.*, **8** (7-8) 635-643 (2007).

[3] R. Riedel (ed.), Handbook of Ceramic Hard Materials, Published by Wiley-VCH, Vol.1 (2000).

[4] T. Wasanapiarnpong, S. Wada, M. Imai, and T. Yano, Effect of Post-sintering Heat-treatment on Thermal and Mechanical Properties of Si$_3$N$_4$ Ceramics with Different Additives, *J. of the Eur. Ceram. Soc.*, **26** 3467-3475 (2006).

[5] T. Y. Tien, Use of Phase Diagrams in the Study of Silicon Nitride Ceramics, *Phase Diagrams in Advanced Ceramics.*, **1** 127-156 (1995).

[6] P. F. Becher, G. S. Painter, N. Shibata, R. L. Satet, M. J. Hoffmann, and S. J. Pennycook, Influence of Additives on Anisotropic Grain Growth in Silicon Nitride Ceramics, *Mater. Sci. and Eng. A.*, **422** 85-91 (2006).

[7] T. Ohji, Microstructural Design and Mechanical Properties of Porous Silicon Nitride Ceramics, *Mater. Sci. and Eng. A.*, **498** 5-11 (2008).

[8] L. Bai, X. Y. Zhao, C. C. Ge, Sintering of β-Si$_3$N$_4$ Powder Prepared by Self Propagating High-Temperatutre Synthesis (SHS), *Mater. Sci. For.*, **546-549** 2179-2182 (2007)

[9] F. Chen, Q. Shen, F. Yan, L. Zhang, Spark Plasma Sintering of α-Si$_3$N$_4$ Ceramics with MgO-AlPO$_4$ as Sintering Adittives, *Mater. Chem. and Physics.*, **107** 67-71 (2008).

[10] C. C. Liu, Microstructural Characterization of Gas-Pressure-Sintered α-Si$_3$N$_4$ Containing β-phase Seeds, *Ceramic International.*, **29** 841-846 (2003).

[11] M. Belmonte, A. de Pablos, M. I. Osendi, and P. Miranzo, Effects of Seeding and Amounts of Y$_2$O$_3$:Al$_2$O$_3$ Additives on Grain Growth in Si$_3$N$_4$ Ceramics., *Mater. Sci. and Eng. A.*, **475** 185-189 (2008).

[12] J. C. Bressianni, V. Izhevskyi, and A. H. A. Bressianni, Development of the Microstructure of the Silicon Nitride Based Ceramics., *Mater. Res.*, **2**, 3 165-172 (1999).

[13] C. Kawai and A. Yamakawa, Crystal Growth of Silicon Nitride Whiskers Through a VLS Mechanism Using SiO$_2$-Al$_2$O$_3$-Y$_2$O$_3$ Oxides as Liquid Phase, *Ceramic International.*, **24** 135-138 (1998).

[14] M. I. Jones, M. C. Valecillos, K. Hirao, and Y. Yamauchi, Grain Growth in Microwave Sintered Si$_3$N$_4$ Ceramics From Different Starting Powders., *J. Eur. Ceram Soc.*, **22** 2981-2988 (2002).

[15] A.C.S. Coutinho, J.C. Bressiani, and A. H. A. Bressianni, Influence of Microstructural Characteristics on the Mechanical Properties of Silicon Nitride with Al$_2$O$_3$, Y$_2$O$_3$ and Nd$_2$O$_3$ as Sintering Aids, *Mater. Sci. For.*, **416-418** 567-572 (2003).

[16] X. Liu, X. Ning, W. Xu, H. Zhou and K. Chen, Study on Thermal Conductivity of Spark-Plasma-Sintered Silicon Nitride Ceramics, *Key Eng. Mater.*, **280-283** 1259-1262 (2005).

[17] G. Ling and H. Yang, Pressureless Sintering of Silicon Nitride with Magnesia and Yttria., *Mater. Chem. and Physics.*, **90** 31-34 (2005).

SUPPRESSION OF SINTERING DEFECTS IN METAL/CERAMIC GRADED LAYERS BY USING INHOMOGENEOUS POWDER MIXTURES

K. Shinagawa[1], and Y. Sakane[2]
[1]Faculty of Engineering, Kagawa University, Takamatsu, Japan.
[2]Graduate School of Engineering, Kagawa University, Takamatsu, Japan.

ABSTRACT
Improvement of the drop in shrinkage rate of metal/ceramic powder mixtures may be important to avoid defects, such as distortion or cracking, in sintering of functionally graded materials. The authors previously proposed a method of increasing the sinterability of metal/ceramic powder mixtures by inhomogenization of microstructure, which was achieved by granulating ceramic and metal powders separately before compaction. In this study, Ni/Al$_2$O$_3$ graded powder compacts are prepared with the inhomogeneous mixtures on some conditions, and the sintering behavior of the compacts is examined analytically and experimentally. Consequently, the effects of inhomogenization on suppressing the sintering defects in graded layers are confirmed.

INTRODUCTION
Typical functionally graded materials (FGM) are laminated composites of tough metal and refractory ceramic, where the volume ratio of ceramic to metal changes from one side to the other, to join the two different materials and also to relieve the thermal stress[1, 2]. When the laminates are fabricated by powder processing, distortion and cracking during sintering often become a problem. To examine the sintering behavior of laminates, analytical methods have been developed for multilayer ceramics[3] and graded layers[4-6]. However, an approach to suppressing the sintering defects on the bases of theoretical analysis is few[7, 8].

Distortions and cracking of metal/ceramic graded layers during sintering may be caused by the variation in sinterability with mixing different types of powder. Especially, adding ceramic particles into metallic powder matrix often causes the significant drop in sintering rate. Improvement of the drop in sinterability of the powder mixtures may be important to avoid the sintering defects in graded layers. In the previous study, the authors proposed a method of increasing the sinterability of metal/ceramic powder mixtures by inhomogenization of microstructure, which was achieved by granulating ceramic and metal powders separately before compaction[9]. The inhomogeneous mixtures, however, have not been practically used to compose multilayer powder compacts yet. In this study, Ni/Al$_2$O$_3$ graded powder compacts are prepared with using the homogeneous (ordinal) or the inhomogeneous mixtures on some conditions, and the difference in sintering behavior between them is examined by numerical analysis as well as experiment. Before numerical analysis, the sintering properties of the two types of mixtures are measured, and expressed by approximate equations.

SINTERING PROPERTIES OF MIXTURES
Preparation of specimens
Ni powder (4SP-400, NOVAMET, mean size of 12.5μm) and Al$_2$O$_3$ powder (AKP53, Sumitomo Chemical Co., Ltd., mean size of 0.3μm) are mixed with a binder (Seruna WK-276, Chukyo Yushi Co., Ltd.) and a dispersion agent (Seruna D305, Chukyo Yushi Co., Ltd.). The volume fraction of Al$_2$O$_3$, X in mixtures is changed from 0 to 1 at an interval of 0.2, which is the same as the previous study[7, 8]. Cylindrical powder compacts with the height of about 12.6mm and the diameter of about 4.7mm are prepared by uniaxial pressing at 39MPa and the subsequent cold isostatic pressing (CIPing) at 196MPa. In advance of the powder compaction, the procedures of mixing and granulating are switched to produce the two series of compacts in respect of particle dispersion, that is, Compact A (ordinal): granules of Ni and Al$_2$O$_3$ mixtures are compacted,

Compact B(inhomogenized): pure granules of Ni and Al_2O_3 are mixed and compacted. The granules are made by putting the powder paste through a sieve with mesh of 425μm in a half-dried state.

Sinter-compression tests

To measure the sintering properties of Compact A and B, the sinter-compression tests at elevated constant temperatures are performed by using a thermomechanical analyzer. The specimens are heated up at a rate of 20K/min in Ar-20%H_2, and the initial applied stress of 0.006MPa are held or changed to σ_z =0.006 - 0.3MPa, which is 5min after reaching prescribed temperatures. The change in height of the specimens is measured for 30min. The strain rate obtained from the reduction in height of the specimens can be expressed by[10]

$$\dot{\varepsilon}_z = \frac{\sigma_z + \Sigma^s}{E} \tag{1}$$

$$E = E_0 \left(t + t_0 \right)^q \tag{2}$$

$$\Sigma^s = (1 - 2v)\sigma^s \tag{3}$$

where E_0 is the longitudinal viscosity for the time t =1 - t_0, v is the viscous Poison's ratio, and σ_s is the sintering stress. t_0 and q are constants, determined by assuming to be common to all the specimens at the same test temperature, and v cannot be evaluated in the used apparatus. Therefore, Σ^s and E_0 are employed for discussing the variations in the sintering stress and the viscosity with mixing ratio X.

Results of measurement

Figure 1 shows the sintering strain rate of each compact at 1223 and 1356K. The sintering strain rate of Compact B with X=0.2 was higher than that of Compact A at both temperatures. The drop in the sintering rate around X=0.2 at 1356K was not large for Compact B, compared with Compact A. Although the relative densities of Compact B were initially lower than those of Compact A, they reached the same level as Compact A around 1473K[9].

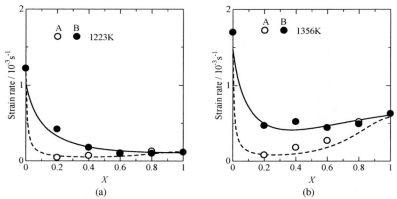

Figure 1. Variation in sintering strain rates with mixing ratio X, expressed by approximate equations
(Symbols: experimental data)

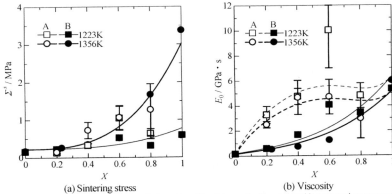

(a) Sintering stress   (b) Viscosity

Figure 2.   Variations in sintering stress and viscosity with mixing ratio X, expressed by approximate equations
(Symbols: experimental data)

Σ^s and E_0 are displayed in Fig. 2.   There was no significant difference in the sintering stress between
Compact A and B.   As for E_0 of mixtures, however, the data on Compact A was much higher than that of
Compact B.   Compact B can be regarded as a composite of the powder compacts of $X=0$ and $X=1$, and a base
of modeling, because its sintering properties were within the range computed by the micromechanics model[11].
Thus the increase in viscosity of Compact A seems to be out of the law of mixtures, but the Al_2O_3 particles
were well dispersed in Compact A in comparison with Compact B[9].   Therefore, the viscosity of the
agglomeration of Al_2O_3 particles, as a component of mixtures, is considered to increase by dispersion, from that
of the original compacts of $X=1$.

Approximate equations
         For conducting the numerical analysis, the sintering properties are expressed by the following
equations;

Sintering stress for Compact A and B:

$$\Sigma^s = 0.544\,X^3 + 0.230 \qquad \text{at 1223K} \tag{4a}$$

$$\Sigma^s = 2.80\,X^3 + 0.224 \qquad \text{at 1356K} \tag{4b}$$

Viscosity for Compact A:

$$E_0 = \alpha\,(4.56\,X^3 - 0.525\,X^2 + 2.12\,X) + 0.196 \qquad \text{at 1223K} \tag{5a}$$

$$E_0 = \alpha\,(3.72\,X^3 - 0.429\,X^2 + 1.73X) + 0.160 \qquad \text{at 1356K} \tag{5b}$$

where

$$\alpha = 1 + 128(H - 0.56)^2 \qquad X \geq 0.56 \tag{6a}$$

$$\alpha = 1 \qquad X < 0.56 \tag{6b}$$

Viscosity for Compact B:

$$E_0 = 4.56\,X^3 - 0.525\,X^2 + 2.12\,X + 0.196 \qquad \text{at 1223K} \qquad (7a)$$

$$E_0 = 3.72\,X^3 - 0.429\,X^2 + 1.73X + 0.160 \qquad \text{at 1356K} \qquad (7a)$$

Since the experimental data for E_0 at low temperature had large errors, the approximate equations for the viscosity at 1223K were roughly estimated from the data of the upper temperatures, as shown in Fig. 3.

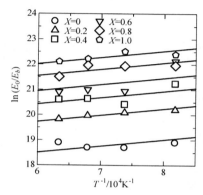

Figure 3.   Relationship between viscosity E_0 and temperature T (E_b=1 Pa·s)

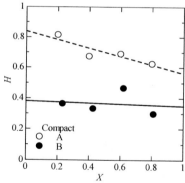

Figure 4.   Mixture homogeneity H versus mixing ratio X
(Symbols: experimental data[8], Lines: Eq. (8))

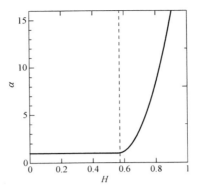

Figure 5. α in Eq. (6), as a function of mixture homogeneity H

The viscosity of agglomeration of Al_2O_3 particles in Compact A was assumed to increase with the degree of dispersion, as referred to above, that is, E_0 in Eq. (5) was set as a function of the mixtures homogeneity H through Eq. (6).   The compacts used in the present study can be regarded as a kind of bimodal powder mixtures because the difference in size between metal and ceramic powder particles is large.   The homogeneity H of the compacts, depending on the degree of dispersion, was evaluated from the theory of packing density of bimodal powder mixtures proposed by German[12], as shown in Fig. 4.   Note that $H=0$ represents inhomogeneous structure due to the perfect separation, and $H=1$ corresponds to homogeneous structure because of the perfect dispersion.

The homogeneity of Compact A was high over the range of mixing ratio, compared with Compact B. In the present study, H was approximated by

$$\text{Compact A: } H = 0.56 + 0.28(1\text{-}X) \tag{8a}$$

$$\text{Compact B: } H = 0.35 + 0.033(1\text{-}X) \tag{8b}$$

The factor α in Eq. (5) was expressed by Eq. (6), as a function of H, which is demonstrated in Fig. 5.

STRESS ANALYSIS OF GRADED LAYERS
Analytical conditions

Methodology for suppressing the sintering defects with the inhomogenized mixtures is sought out theoretically.   Surface cracking may be caused by the tensile stress on the top ceramic layer, which stems from the mismatch shrinkage as well as the bending of multilayer.   Since the radial and the circumferential stresses are the same, and constant along the radial directions in the disc specimens except the periphery[7], the distribution of radial stress along the center axis in graded layers is observed, and the geometrically linear plate theory[5] is used for ease instead of the finite element method in the same way as the previous study[8], with Eqs. (4) - (7), where ν is set to be 0.3.   The graded structures with six layers are treated in the analysis, as indicated in Table 1, where the six layers correspond to the compacts of $X=0$, 0.2, 0.4, 0.6, 0.8, 1.

The different structures are considered to separately examine the effects of inhomogenization of mixtures, the size of powder particles in F1 and F6 layers, and the thickness of F6 layer on the stress distribution. The sintering properties of Ni and Al_2O_3 powder with large particles ($F1_m$ and $F6_m$) used in the analysis are shown in Table 2, where $F1_m$ and $F6_m$ correspond to NIE05PB (<150μm, Kojundo Chemical Lab Co., ltd.) and AKP20 (0.57μm, Sumitomo Chemical Co., Ltd.), respectively. Since the sintering properties change with temperature, the analysis is performed for both case of 1223K and 1356K. The basic thickness of each layer is taken to be 1.2 mm. Thus the thickness ratio of 1/4 in Table 1 represents 0.3mm.

Table 1. Graded Structures

Graded layers	Components bottom.........top	Thickness ratio F1~F5 : F6	Mixtures
GS-A	F1, F2, F3, F4, F5, F6	1 : 1	Ordinal
GS-B	F1, F2, F3, F4, F5, F6	1 : 1	Inhomogenized
GS-B_m	$F1_m$, F2, F3, F4, F5, $F6_m$	1 : 1	Inhomogenized
GS-B_q	F1, F2, F3, F4, F5, F6	1 : 1/4	Inhomogenized
GS-A_{mq}	$F1_m$, F2, F3, F4, F5, $F6_m$	1 : 1/4	Ordinal
GS-B_{mq}	$F1_m$, F2, F3, F4, F5, $F6_m$	1 : 1/4	Inhomogenized

Table 2. Sintering Properties of Ni($F1_m$) and Al_2O_3($F6_m$) Powders

Powder	1223K		1356K	
	Σ^s / MPa	E_0 / GPa•s	Σ^s / MPa	E_0 / GPa•s
$F1_m$	0	0.0100	0	0.0349
$F6_m$	0.315	3.54	0.793	1.13

Analytical results and discussion

Figure 6 shows the distribution of the radial stress along center axis in the layers, where all have fundamental comparisons between GS-A, GS-B and GS-B_{mq}. The magnitude of the tensile stress in 100% Al_2O_3 layer (X=1), i.e. F6(z = 6.0-7.2mm) or $F6_m$ (z = 6.0-6.3mm), is focused on this study, because, in general, ceramic layer is more brittle than the others. As can be seen in Fig. 1, the sintering balance of layers, except F1(X=0), is good for Compact A at 1223K, and also for Compact B at 1356. Corresponding to the well balanced sintering rates, the stress distribution in each layer is flat for GS-A at 1223K, but for GS-B at 1356K. Reversely, although the tensile stress in 100%Al_2O_3 layer for GS-A becomes large at 1256K, that for GS-B is also large at 1223K. On the other hand, the tensile stress in GS-B_{mq} is lower than those in GS-A and GS-B for both temperatures. It is confirmed that using $F1_m$ and $F6_m$ or thinning F6 layer alone is not effective, as shown for the case of GS-B_m and GS-B_q, where the tensile stress cannot be reduced. The necessity of the inhomogenized layers (Compact B) is also verified by comparison with GS-A_{mq}, in which the tensile stress is not reduced much at 1356K. Therefore, the combination of three factors, particle size, layered structure, and inhomogeneous mixtures may be needed to suppress the tensile stress generated in 100%Al_2O_3 layer up to 1356K.

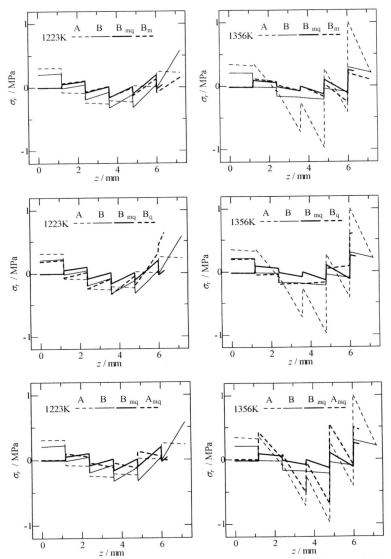

Figure 6.   Distribution of radial stress along center axis in graded layers, GS-A, B, B_{mq} and others

SINTERING EXPERIMENT OF GRADED LAYERS

Experimental conditions

GS-A, GS-B and GS-B$_{mq}$ are actually fabricated, and the surface cracking is examined in the sintering experiment.   The graded layers are pressed into discs with a diameter of 40mm and 1.2mm thickness of each layer, under a pressure of 39MPa, and compacted by CIPing of 196MPa.   The ratio of diameter to thickness may be enough to produce the region where the stress distribution is constant apart from the edge of discs.   The size of discs in the present range may also not affect the magnitude of the stress, according to the geometrically linear plate theory, while the warping should be normalized by the diameter of discs (see Fig. 7).   The specimens are heated up at a rate of 100K/h in Ar-3%H$_2$, kept at 1223K, 1323K, or 1423K for 10min, and cooled down at a rate of 100K/h.   Sintering at 1623K for 120min is also completed in the same way, for reference.   After cooling, distortion and cracking in the graded layers are checked.

Experimental results and discussion

Figure 7 shows the deformation behavior of the specimens.   The specimens tended to warp to the side of Ni under 1323K because the shrinkage rate of 100%Ni layer (X=0) was much larger than that of the others in the initial stage of sintering.   At the later stage, the specimens turned to the side of Al$_2$O$_3$ due to the shrinkage of 100%Al$_2$O$_3$ layer (X=1) at high temperature.   The tendency of warping shows good agreement with the stress distribution obtained in the analysis, represented in Fig. 6.   Note that the warping to the side of Ni may produce the bending tensile stress in 100%Al$_2$O$_3$ layer, which has a possibility of surface cracking.

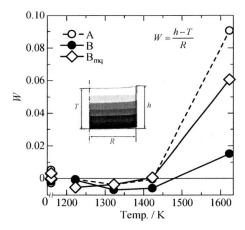

Figure 7. Warping of graded layers, GS-A, B and B$_{mq}$

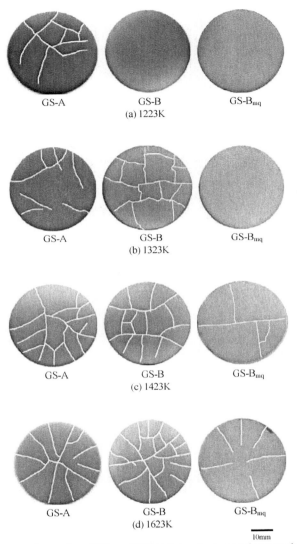

GS-A GS-B GS-B$_{mq}$
(a) 1223K

GS-A GS-B GS-B$_{mq}$
(b) 1323K

GS-A GS-B GS-B$_{mq}$
(c) 1423K

GS-A GS-B GS-B$_{mq}$
(d) 1623K

10mm

Figure 8.   Appearance of top surface of F6 layer (100%Al$_2$O$_3$ layer), where cracks are emphasized by drawing

Figure 8 displays the appearance of the top surface of 100%Al_2O_3 layer in each specimen. There was no cracking in GS-B_{mq} up to 1323K. This agrees well with the low tensile stress in the analytical results for GS-B_{mq}, as can be seen in Fig. 6. Since the initial relative density of F6$_m$, used in GS-B_{mq}, is similar to that of F6 (about 0.61)[7], and not increased much below 1356K, the strength of the ceramic layer may not be so different among the specimens. Thus the experimental results may also support the effectiveness of the use of inhomogenized mixtures in the graded layers. Cracking above 1423K may be mainly caused by the stress due to the mismatch shrinkage between 100%Al_2O_3 layer and the substrate. This type of defect is not a subject of the present study, but should be solved elsewhere.

CONCLUSIONS

In the Ni/Al_2O_3 graded compact with homogenous powder mixtures, made by an ordinary procedure, surface cracking occurred on 100%Al_2O_3 layer at the early stage of sintering due to the poor sintering balance among the layers. The effects of the inhomogenization of powder mixtures for suppressing the cracks were examined by numerical analysis, where the variations in sintering stress and viscosity measured by sinter-compression tests were modeled, and taken into consideration. Consequently, it was found that the tensile stress in 100%Al_2O_3 layer can be reduced by using the inhomogenized powder mixtures, in combination with adjusting the particle size and the layer thickness. The sintering experiment of graded layers were also conducted, and it was confirmed that the surface cracking at the early stage of sintering did not occur actually in the above condition. Thus the effectiveness of the inhomogenization of microstructure, as a way of suppressing sintering defects in metal/ceramic graded powder compacts, was verified.

ACKNOWLEDGMENT
This work was supported by JSPS.KAKENHI(19560733).

REFERENCES
[1]S. Suresh and A. Mortensen, *Fundamentals of Functionally Graded Materials*, (1998), IOM Communications.
[2]Y. Miyamoto, W. A. Kaysser, B. H. Rabin, A. Kawasaki and R. G. Ford, *Functionally Graded Materials: Design, Processing and Applications*, (1999), Kap.
[3]P. Z. Cai, D. J. Green and G. L. Messing, Constrained Densification of Alumina/Zirconia Hybrid Laminates, II: Viscoelastic Stress Computation, *J. Am. Ceram. Soc.*, **80**-8, 1940-1948 (1997).
[4]K. Shinagawa: Deformation Analysis of Graded Powder Compacts during Sintering, in *Functionally Graded Materials 1996*, Eds. by I. Shiota and M. Y. Miyamoto, 69-74 (1997), Elsevier.
[5]H. Riedel and T. Kraft, Distortions and Cracking of Graded Components during Sintering, *Materials Science Forum*, **308-311**, 1035-1040 (1999).
[6]M. Gasik and B. Zhang, Sintering of FGM Hardmetals in Different Conditions: Simulation and Experimental Results, *Ceram. Trans.*, **114**, 341-347(2001).
[7]K. Shinagawa and Y. Hirashima: Viscoplastic Stress Analysis of Shrinkage and Warpage in Graded Layers during Sintering, *Key Engineering Materials*, **233-236**, 785-790 (2003).
[8]K. Shinagawa and Y. Hirashima, Stress Analysis of Powder Compacts with Graded Structures in Sintering Process, *Mat. Sci. Forum* **492-493**, 477-82 (2005).
[9]K. Shinagawa, Effects of Inhomogenization on Sintering Behavior on Ni/Al_2O_3 Powder Mixtures, *Mat. Sci. Forum*, **631-632**, 245-50(2010).
[10]K. Shinagawa, Sintering Stress and Viscosity of Ni/Al_2O_3 Powder Mixtures, in *Multiscale and Functionally Graded Materials*, Eds. by G. H. Paulino et al., 22-27(2008), AIP.
[11]K. Shinagawa: Variations in Sintering Stress and Viscosity with Mixing Ratio of Metal/Ceramic Powders,

Ceramic Transactions, **209**, 161-70 (2010).
[12]R. M. German, Prediction of Sintered Density for Bimodal Powder mixtures, *Metall. Trans.*, **23A**, 1455-65 (1992).

CO-SINTERING OF AN ANODE-SUPPORTED SOFC BASED ON SCANDIA STABILIZED ZIRCONIA ELECTROLYTE

T. Reynier[a, b], D. Bouvard[a], C.P. Carry[a], R. Laucournet[b]

[a]Laboratoire SIMAP, Grenoble INP / CNRS / UJF, BP46, 38402 Saint Martin d'Hères, France
[b]CEA-LITEN, 17 rue des Martyrs, 38054 Grenoble, France

ABSTRACT
    This paper investigates the sintering of an anode supported SOFC cell based on scandia and ceria stabilized zirconia electrolyte. We focus on the co-sintering of anode-electrolyte half cell. At first, the sintering of electrolyte and NiO/YSZ anode has been characterized by conventional dilatometry experiments. Then, the sintering of a half cell has been performed by direct observation of bending occurring during the thermal cycle. Finally, the observed cambering sequences have been discussed by Cai et al. analytical model.

INTRODUCTION

    For decades, Solid Oxide Fuel Cells (SOFC) have been recognized as attractive energy conversion devices[1]. Currently, an operating temperature of 800°C is necessary for energy conversion. This high temperature limits the lifetime of the system due to materials damage. The reduction of this temperature is the primary area of research in SOFC[2, 3]. One of most reasonable solutions is the use of materials with improved electrochemical properties. For example, using Scandia and ceria Stabilized Zirconia (SSZ) instead of classical yttria stabilized zirconia (YSZ) as electrolyte significantly improves the cell performances[4].

    Another issue of SOFCs is their high cost, particularly due to the number of manufacturing process steps, which generally includes two to three sintering operations. For example, Jülich Forschungzentrum process[5] comprises a pre-sintering of anode support at 1250°C, followed by a sintering of the functional layer and the electrolyte at 1400°C and to finish the cathode sintering at 1100°C. Other groups proposed two successive stages: co-sintering of anode-electrolyte stacking and cathode sintering[6]. For sure, the development of a new process allowing the sintering of a complete cell in one single step would lead to a significant reduction of the global cells cost.

    A critical aspect of co-sintering is the distorsion induced by the shrinkage mismatches between the materials constituting the cell. It is difficult to get rid of these mismatches because the electrodes should be highly porous whereas the electrolyte should be dense. Several authors already investigated this question. For example, Mücke et al.[7] worked on the co-sintering of anode support (NiO/8YSZ) and electrolyte (8YSZ) and found that half-cell bent first towards one side, got flat again and then bent in the other way. The authors correlated this behaviour with the strain rate of each layer and concluded that a flat half-cell could be obtained by limiting the duration of the isothermal period. Lee et al.[8] have been able to co-sinter a NiO-YSZ/YSZ/LSM cell without curvature by applying a low load (~ 0.1kPa) at a particular time of the sintering, when the materials presented the lowest viscosities.

    The present paper reports the development of a new anode-support SOFC cell including SSZ electrolyte and processed with a single sintering step. It focuses on the problem of distorsion during co-sintering of half cell. First, both electrolyte and anode materials have been characterized by conventional dilatometry measurements. Half-cells have then been fabricated by anode tape casting followed by electrolyte screen printing. The deformation of these components during sintering has been continuously observed by optical dilatometry. Furthermore, this deformation has been compared with the prediction of Cai et al.[9] analytical model.

EXPERIMENTAL PROCEDURE

The sintering of SSZ ($10mol\%Sc_2O_3$–$1mol\%CeO_2$–ZrO_2) powder and NiO/8YSZ ($NiO/8mol\%Y_2O_3$-ZrO_2) powder mixture has been investigated by dilatometry measurement performed with a Setaram TMA 92 device under flowing air atmosphere. 10Sc1CeSZ, NiO and 8YSZ powders have been provided respectively by DKKK, Nowamet and Tosoh companies. Their specific surface areas are 11 m^2/g, 3.5 m^2/g and 6.5 m^2/g, which correspond to equivalent particle diameters of 90 nm, 260 nm and 150 nm, respectively. Dilatometry experiments have been achieved with compacts of SSZ and NiO/8YSZ powders. In the fabrication process, these powders will be shaped by screen printing and tape casting, with densities estimated to 0.40 and 0.47, respectively. In order to determine the sintering behaviour of both components, compacts with the corresponding green densities have been used. Thus, a pressure of 70 MPa is required to get a 40% dense SSZ compact and 45 MPa to get a 47% dense NiO/8YSZ compact (□□□□8 mm, h = 10 mm).

For the preliminary co-sintering experiments, 3 cm diameter half cells have been fabricated in two steps. At first, the anode support was prepared by tape casting. The slurry was composed of 50wt.%NiO/50wt.%8YSZ - ethanol / methyl-acetone (solvents) - CP213 (dispersant) - PEG 400 (plasticizer) - PVB90 (binder). Several stages of milling in planetary mixer, followed by a deaeration, provided the slurry, which was tape casted with a thickness of 1000 µm, decreasing to 500 µm after drying. Next, the electrolyte powder was deposited by screen printing on the green anode support. SSZ ink was prepared by a mixture of terpineol and 5 wt% ethyl cellulose and homogenized by milling.

The in situ observation of sintering was performed with TOM-AC optical dilatometer designed by ISC Fraunhofer with molybdenum-wire heating elements under argon atmosphere. The samples used for these experiments were 4 cm long and 1 cm wide bilayer parts, with an anode thickness of 500 µm and 20 µm electrolyte.

MATERIALS AND SINTERING BEHAVIOURS CHARACTERIZATIONS

The electrolyte is an ionic conductor. It transports O^{2-} ions from the cathode to the anode, thanks to oxygen vacancies in the crystalline structure of zirconia doped with scandium (Sc^{3+}). The cathode material, not discussed in this paper, is a mixed ionic and electronic conductor Nd_2NiO_4, and known as a promising and recently developed material[10]. This material reduces $O_2(g)$ into O^{2-} anions. The anode material is a cermet Ni/8YSZ. It permits the reaction of H_2 oxidation at the Triple Phase (Nickel-8YSZ-gaz) Boundaries (TPBs). The anode is fabricated from a powder mixture of zirconia and nickel oxide. The NiO/8YSZ composite obtained after sintering is then reduced in situ by H_2 in the preheating step of cell. This reduction is accompanied by a 40% volume decrease of particles containing nickel and gives the final anode porosity. For example, 20% porosity is induced by the reduction of nickel oxide particles in a 50vol.%NiO/50vol.% 8YSZ composite. Thus, it is possible to sinter the anode at a relatively high temperature without the risk of obtaining a too low porosity after reduction.

Previous studies have permitted to set the optimal sintering conditions for the complete cell. The criteria have been a fully dense electrolyte and a cathode with at least 30% porosity. It has been found that the sintering conditions 1200°C-6h, 1250°C-4h, 1300°C-3h and 1350°C-3h allowed obtaining the SSZ electrolyte with closed porosity. Ionic conductivity measurements were performed on sintered SSZ samples. A conductivity of 0.048 $S.cm^{-1}$ was obtained at 700°C for each sample. This value is consistent with literature values[3,4] and is much higher than the 8YSZ ionic conductivity (<0.02S.cm^{-1} at 700°C).

Figures 1 and 2 show the conventional dilatometry measurements performed on anode and electrolyte compacts during the following thermal cycle: 5°C/min heating and 3 hour holding at 1350°C. SSZ material sintered with a strain rate peak at 1170°C. A linear shrinkage of about 26% is required to achieve densification with a closed porosity. NiO/YSZ composite presents a lower linear shrinkage of about 20% after sintering. This material exhibits a strain rate peak at 1325°C.

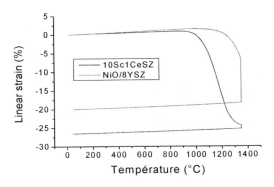

Figure 1. SSZ and NiO/8YSZ linear shrinkage. Sintering at 1350°C-3h
with a 5°C/min heating rate.

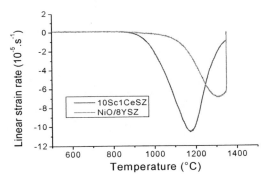

Figure 2. SSZ and NiO/8YSZ linear strain rates. Sintering at 1350°C-3h
with a 5°C/min heating rate.

CO-SINTERING EXPERIMENTS

First co-sintering experiments were conducted on an anode/electrolyte half-cell. Various thermal cycles defined in the previous section have been tested. Samples cross section observations by SEM revealed an open porosity in the electrolyte for sintering temperatures lower than 1350°C. The porosity decreases with increasing sintering temperature. This result was attributed to the mismatch between the shrinkage of both materials. Indeed, in co-sintering approach, the linear electrolyte shrinkage is limited to 20% in the horizontal directions, i.e. the anode shrinkage, while the material requires, in the case of free sintering, a linear shrinkage of 26% to get a closed porosity. The electrolyte densification can thus only be explained by an anisotropic shrinkage, that is to say, a larger vertical shrinkage to compensate the 20% shrinkage in the horizontal directions. Finally, it appeared that a 3 hours dwelling at 1350°C is necessary to obtain a sufficiently dense electrolyte (Figure 3). Therefore, these conditions were chosen for the half cell co-sintering.

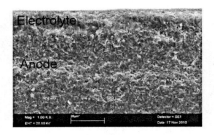

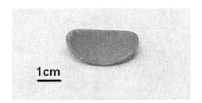

Figure 3. SSZ electrolyte and NiO/8YSZ anode co-sintered at 1350°C-3h

Figure 4. Half cell photography after co-sintering at 1350°C-3h. Curvature towards (green) anode side

It was also observed that all samples presented a concave curvature toward the anode side (Figure 4). The curvature did not depend on which face (anode or electrolyte) was initially set on the support. This observation is counter-intuitive because the electrolyte shrinkage is supposed to be higher than the anode shrinkage.

OBSERVATION OF CO-SINTERING
The sintering of half-cell samples has been continuously observed by the optical dilatometer. The rectangular samples were placed with the electrolyte facing the alumina support (Figure 5).

Figure 5. Half-cell electrolyte/anode sample on an alumina support in the optical dilatometer at room temperature

The experiment was performed under argon at 1350°C with a 2h dwelling time and a heating rate of 20°C/min. We observed a complex sequence, which is described below.

- Between 1050 and 1200°C: first curvature towards electrolyte side with very limited planar shrinkage. (Figure 6)

Figure 6. Half cell at 1150°C – curvature towards electrolyte face

- Between 1200°C and during the first half of dwelling time (1350°C, 1h) : global planar shrinkage and sample flattening (Figure 7).

Figure 7. Half cell at 1350°C+1h. Planar shrinkage and sample flattening.

- During the second half of the dwelling time: curvature towards the anode side (Figure 8). This curvature is also met after cooling.

Figure 8. Half cell at 1350°C+2h. Curvature towards anode side.

ANALYTICAL MODELLING

The sintering behaviour of the half cell has been modelled to be better understood. According to the analytical model proposed by Cai et al.[9], the radius of bilayer curvature depends on the thickness ratio, viscosity ratio and sintering shrinkages. The normalized curvature is defined as the curvature multiplied by the total thickness of the bilayer sample. The normalized curvature rate during co-sintering is thus expressed as:

$$\dot{k} = \frac{h_1 + h_2}{\dot{r}} = \frac{6(m+1)^2 mn}{m^4 n^2 + 2mn(2m^2 + 3m + 2) + 1} \Delta\dot{\varepsilon} \qquad (1)$$

with $m = \dfrac{h_1}{h_2}$ and $n = \dfrac{\eta_1/(1-\upsilon_1)}{\eta_2/(1-\upsilon_2)}$, where h_1 and h_2 are the thicknesses of each layer, η_1 and η_2 are the viscosities of each material, v_1 and v_2 are their viscous Poisson's coefficients and $\Delta\dot{\varepsilon} = \dot{\varepsilon}_2 - \dot{\varepsilon}_1$ is the differential strain rate between both layers. The curvature rate is assumed to be proportional to $\Delta\dot{\varepsilon}$, which can be easily obtained by subtraction of NiO/YSZ and SSZ strain rates measured by dilatometry at the same time during the same thermal cycle (Figure 9).

First the differential strain rate is positive, when the electrolyte sinters faster than the anode and then becomes negative when it is the opposite. m parameter is calculated as 20/300. The viscous Poisson's coefficient is obtained by the Bordia and Scherer relation[11].

$$v^{\upsilon p} = 0.5\sqrt{\frac{\rho}{3 - 2\rho}} \qquad (2)$$

where ρ is the relative density calculated from the conventional dilatometry measurements.

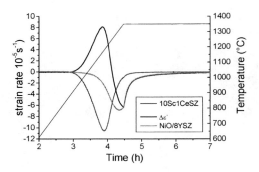

Figure 9. Differential strain rate of anode and electrolyte layers, $\Delta\dot{\varepsilon} = \dot{\varepsilon}_{NiO/YSZ} - \dot{\varepsilon}_{SSZ}$.
Sintering at 1350°C-3h, heating rate 5°C/min.

The materials viscosities during the sintering cycle have been estimated from three point bending experiments during an identical sintering cycle[12]. The obtained values are given in Figure 10.

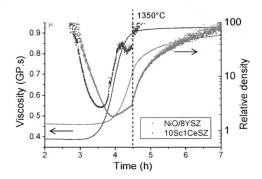

Figure 10. SSZ and NiO/YSZ viscosities and relative densities versus time during the sintering cycle: heating rate 5°C/min up to 1350°C + 3 h at 1350°C.

 Subsequently, the normalized curvature rate has been calculated from Eq. 1 for both materials and drawn in Figure 11. As expected, it follows the same trend as the differential strain rate. However, at the beginning of sintering, the increase of the curvature rate curve is delayed with regard to the increase of the differential strain rate. This is due to the high anode viscosity compared to the electrolyte viscosity in this temperature range. Because of this viscosity ratio and the thickness ratio (electrolyte 20 μm / anode 300 μm) as well, the anode is too strong to be deformed by the electrolyte deformation and thus the cambering phenomenon is retarded.

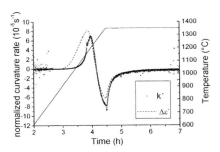

Figure 11. Normalized curvature rate of the SSZ-NiO/YSZ bilayer during sintering at 1350°C-3h. Heating rate 5°C/min.

Then the normalized curvature has been calculated by time integration of the normalized curvature rate (Figure 12). It finally appears that curvature is first positive, which means a curvature towards electrolyte side, and then negative (curvature towards anode side) until the sintering end. The model predicts a larger curvature towards electrolyte side than towards anode side and the opposite is observed by optical dilatometry. However, the magnitude order of the final curvature (0.02) is in agreement with the sintered bilayer. Calculations are thus in qualitative accordance with previously presented experimental observations.

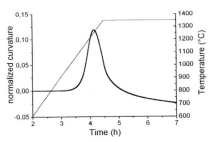

Figure 12. Normalized curvature of the SSZ-NiO/YSZ bilayer during sintering at 1350°C-3h. Heating rate 5°C/min.

DISCUSSION

Optical dilatometry experiments were performed in argon atmosphere. The low oxygen partial pressure certainly induces the reduction of NiO particles located close by the sample surface and may have an impact on the deformation of the half-cell during co-sintering. To check the importance of this phenomenon, we also conducted conventional dilatometry experiments with NiO/YSZ and SSZ compacts in argon. These experiments showed that the sintering strain rate mismatch is almost the same than in air. Therefore we assume in the following that the deformation of the half-cell is not affected by the change in atmosphere.

The cambering of the anode-electrolyte bilayer during sintering is mostly due to the mismatch of the material sintering strain rates $\Delta\dot{\varepsilon}$. In the first part of the co-sintering cycle, below 1200°C, when $\Delta\dot{\varepsilon}$ is positive, a limited densification of the electrolyte upon non-densifying anode occurs, which results in a curvature towards electrolyte side, as soon as the viscosity of the anode is low enough. As the electrolyte is about 15 times thinner than the anode, it could not impose a planar shrinkage.

In the second part of the co-sintering cycle, between 1200°C and the first half of the dwelling time at 1350°C, $\Delta\dot{\varepsilon}$ becomes negative. The anode sintering is booming while the electrolyte finishes its densification, which leads to the sample flattening and a global shrinkage imposed by the anode. During the final stage of the sintering, $\Delta\dot{\varepsilon}$ stays negative. The anode finishes its densification. The electrolyte is dense but it can be deformed due to its low viscosity. Thus we observe an increasing curvature towards anode side until the end of sintering without significant planar shrinkage.

CONCLUSION

We studied the co-sintering of a SSZ/NiO-YSZ half cell. SSZ electrolyte was found to be dense enough after free sintering at 1200 and 1300°C but not after half cell co-sintering because the anode limited the electrolyte planar shrinkage. Nevertheless sufficient densification was obtained after co-sintering at 1350°C due to anisotropic electrolyte deformation, as a higher shrinkage in thickness compensated the reduced planar one. The differences in sintering kinetics between both materials were also investigated to explain the curvature observed at the end of half cell co-sintering. Continuous observation of the sintering process showed that the half cell bilayer presented two successive curvatures of the bilayer during sintering. This behaviour has been explained by an analytical model mainly based on the differential strain rate $\Delta\dot{\varepsilon}$. We found in particular that a thin electrolyte, 15 times thinner than the anode layer in our case, was able to bend the bilayer sample at the beginning of sintering and that the global shrinkage was controlled by the thickest component. With materials selected in this study, it seems inconceivable to obtain a flat cell without applying a load. The next challenge will be to sinter a fully cell including a porous cathode and to get it flat with safe interfaces.

ACKNOWLEDGMENTS

The authors thank Rhône-Alpes region and Energies Research Cluster for funding T. Reynier PhD. thesis.

REFERENCES

[1] P. Timakul, S. Jinwath, P. Aungkavattana., Fabrication of electrolyte materials for solid oxide fuel cells by tape-casting, *Ceramics International*, **34**, 867–871 (2008).

[2] H. Moon, S.D. Kim, S.D. Hyun, and H.S. Kim, Development of IT-SOFC Unit Cells with Anode-Supported Thin Electrolytes Via Tape Casting and Co-firing, *Inter. J. Hydrogen Energy*, **33**, 1758-1768 (2008).

[3] M.A. Laguna-Bercero, S.J. Skinner, J.A. Kilner, Performance of solid oxide electrolysis cells based on scandia stabilised zirconia, *J. Power Sources*, **192** 126-131 (2009).

[4] D.S. Lee, W.S. Kim, S.H. Choi, J. Kim, H.W. Lee, J.H. Lee, Characterization of ZrO_2 co-doped with Sc_2O_3 and CeO_2 electrolyte for the application of intermediate temperature SOFCs, *Solid State Ionics*, **176** 33-39 (2005).

[5] http://www2.fz-juelich.de/ief/ief-1/index.php?index=66

[6] K.C. Wincewicz, J.S. Cooper, Taxonomies of SOFC material and manufacturing alternatives, *J.Power Sources*, **140**, 280–296 (2005).

[7] R. Mücke, N.H. Menzler, H.P. Buchkremer, D. Stöver, Cofiring of Thin Zirconia Films During SOFC Manufacturing, *J. Am. Ceram. Soc.*, **92** 95-102 (2009).

[8] S.H. Lee, G.L. Messing, M. Awano, Sintering Arches for cosintering Camber-Free SOFC Multalayers, *J. Am. Ceram. Soc.*, **91** 421-427 (2008).

[9] P.Z. Cai, D.J. Green, G. Messing, Constrained Densification of Alumina/Zirconia hybrid Laminates, II : Viscoelastic Stres Computation, *J. Am. Ceram. Soc.*, **80** 1940–1948 (1997).

[10]F. Chauveau, J. Mougin, J.M. Bassat, F. Mauvy, J.C. Grenier, A new anode material for solid oxide electrolyser: The neodymiumnickelate $Nd_2NiO_{4+\delta}$, *J.Power Sources*, **195**, 744–749 (2010).

[11]R.K. Bordia, G.W. Scherer, On constrained sintering-II, Comparison of Constitutive Models, *Acta Metall.*, **36**, 2399-2409 (1988).

[12]HG. Kim et al., A phenomenological constitutive model for the sintering of alumina powder, *J. Eur. Ceram. Soc.*, **23**, 1675–1685 (2003).

BULK DOPING INFLUENCE ON GRAIN SIZE AND RESPONSE OF CONDUCTOMETRIC
SnO₂-BASED GAS SENSORS: A SHORT SURVEY

G. Korotcenkov[a] and B.K. Cho[a,b]

[a]Department of Material Science and Engineering, Gwangju Institute of Science and Technology
Gwangju, Republic of Korea (ghkoro@yahoo.com)
[b]Department of Nanobio Materials and Electronics, Gwangju Institute of Science and Technology
Gwangju, Republic of Korea (chobk@gist.ac.kr)

ABSTRACT
    The influence of bulk doping by various additives on both the grain size and the response of
SnO₂-based gas sensors is analyzed in this paper. It is shown that during bulk doping of metal oxides
aimed for gas sensor application we should be very careful in selection of doping concentration,
because for attainment of both strong decrease of the grain size and the improvement of thermal
stability we need to use high concentration of additives, while for achievement optimization of gas
sensing characteristics the concentration of additives should not exceed 1-3 %.

INTRODUCTION
    Polycrycrystalline semiconductor oxides such as SnO₂ are widely used in solid state gas
sensors.[1-3] However till now studies in this area are being focused on improvement properties of this
material. It was established that bulk doping of semiconductor metal oxides (MOX) by various
additives is one of the main approaches for resolving a problem, connected with improvement of
sensitivity and selectivity, decreasing the operating temperature, and enhancing the response rate of
gas sensors.[1,2,4-8] Research has shown that additives incorporated in metal oxides even in small
quantity can modify parameters of metal oxides important for achievement desired gas sensing
characteristics such as morphology, electroconductivity, the height of intergrain potential barrier, and
surface reactivity. Doping additives can also stabilize a particular valence state of metal, favor
formation of active phases, stabilize the catalyst against reduction and increase the electron exchange
rate. This influence takes place usually through the change of the concentration of point defects,
mainly oxygen vacancies, the transformation of surface stoichiometry, the appearance of catalytically
active clusters, affecting catalytic reactivity of metal oxide surface, and the change of phase
composition accompanied by segregation of second phase on the surface of the metal oxide grains. As
it is known, morphology controls 3-D conductivity network and gas penetrability of metal oxide
matrix; oxygen vacancies control both bulk and surface properties of metal oxides, in particular, the
chemisorption of oxygen, water and detected gas; catalytic activity controls the rate of surface
reactions including for example CO, H₂ and CH₄ oxidation accompanied by decrease of surface charge
trapped by oxygen species; surface segregation of second phase hinders from the grain growth, etc.[4,9]
    Research has shown that impurities also need to be added to conventional binary MOXs to
stabilize their (meso)porous nanocrystaline morphology.[6] In addition it was found that introduction of
special doping microadditives of various impurities into metal oxides during their synthesis could
change the conditions of the base oxide growth, sufficiently decrease the grain size and improve
thermal stability of the grain size in formed ceramics. Moreover, as it was established in numerous
papers[1,4,7-13] the decrease of the grain size is the simplest way for achievement of better sensor
response.
    At present the influence of the grain size on the conductivity and adsorption and catalytic
properties of metal oxides may be attributed to the fundamentals of operation of conductometric gas
sensor. For description of metal oxide conductivity it is usually used modified "grain" model,[14-16]
according to which the conductivity of 3-D networks of grains is determined by the resistance of

intergrain contacts, which is controlled by the height of Schottky barrier at the inter-grain boundary. Potential barrier is formed due to the oxygen adsorption and the trapping of electrons from conduction band.[10] This effect was observed for WO$_3$, SnO$_2$, In$_2$O$_3$, ITO and many other metal oxide-based gas sensors.[3] It was established that in the big enough range, the decrease of the grain size (t) in metal oxides should be accompanied by the increase of sensor response. According to this model especially strong increase of sensor response is being expected at grain sizes, comparable with Debye length (L_D) or the thickness of depleted layer (L_s). If $t < 2L_s$, where L_s is the width of surface space charge, every grain is fully involved in space charge layer and the electron transport is affected by the charge at adsorbed species.[15,17] In gas sensor the charge at adsorbed species is usually controlled by operating temperature and surface reactions such as adsorption/desorption and catalytic reaction of detected gas with oxygen species presented on the surface. More detailed consideration of gas interaction with metal oxides may be found in several good articles,[9,15,18-21] where different phenomenological models were proposed for description gas sensor operation.

So, bulk doping of metal oxides opens up exciting additional possibilities for varying the structure, electro-physical and catalytic properties of metal oxide materials aimed for gas sensor application.

DOPING INFLUENCE ON THE GRAIN SIZE

At present there are many experimental works devoted to study of doping influence on structural parameters of SnO$_2$. Research has shown that for optimization of the grain size in metal oxide ceramics different additives can be used. Examples of doping influence on the SnO$_2$ grain size are shown in Figure 1.

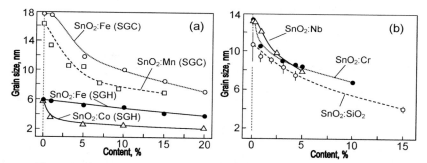

Figure 1. The average crystallite size of the doped SnO$_2$ samples with different content of additives calculated from Scherrer formula: (a) SnO$_2$:Co nanoparticles were prepared by a simple sol–gel-hydrothermal method. SnO$_2$:Fe and SnO$_2$:Mn nanoparticles were prepared by sol–gel calcination (SGC) and sol–gel-hydrothermal routes (SGH), respectively. Data extracted from refs.[27,28]; (b) SnO$_2$:Cr nanoparticles were produced by a polymer precursor method. SnO$_2$-SiO$_2$ powders were synthesized by flame spay pyrolysis technique. Data extracted from refs.[29-32]

It was found that the decrease of the grain size and its stabilization during annealing (see Figure 2), i.e. the inhibition of particles growth during annealing, was observed for SnO$_2$ doped by V,[22] Ce[23] and many other metals and non-metals such as P-Ba, Sm, Ba, P, Mo, W, Ca, Sr, and Cr. For example, research has shown that the impregnated foreign additives (5 at%), consisting of both oxides and polyoxy compounds of indicated elements could keep D less than 10 nm even after calcination at 900 °C, whereas pure SnO$_2$ underwent the grain growth to have D of 13 and 27 nm at 600 and 900 °C, respectively.[1,24] These data for SnO$_2$:Ce, SnO$_2$:La and SnO$_2$:Y are shown in Table 1. Reduced sintering is also reported for indium-doped[15] and antimony doped tin dioxide.[26] Results related to influence of Fe doping on the grain size during annealing are presented in Figure 1a.

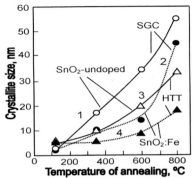

Figure 2. Averaged crystallite size of the SnO$_2$ powders vs. temperature of annealing. Undoped SnO$_2$ (1,3) and Fe-doped SnO$_2$ nanoparticles were prepared by sol–gel calcination (SGC) and sol–gel-hydrothermal routes (HTT). Data extracted from ref.[28]

Table 1. Influence of additives (5%) and annealing temperature on the SnO$_2$ grain size. (Data extracted from ref.[33]

Samples	Average grain size, nm		
	550 °C	900 °C	1100 °C
SnO$_2$ (undoped)	12.7	44.7	158.7
SnO$_2$:Ce	11.7	16.9	70.3
SnO$_2$:La	6.2	8.7	50.4
SnO$_2$:Y	5.2	11.1	47.3

It has been found that mentioned above effects took place due to a surface segregation of the doping elements during the annealing processes (see Figure 3).[34,35] Foreign cations, moved to the particle surface, decrease particle growth velocity.

Grains of second phase

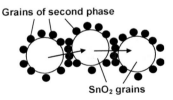

SnO$_2$ grains

Figure 3. Diagram illustrating the segregation of the second phase on the surface of the SnO$_2$ grains.

The surface segregation layers can strongly affect the properties of the materials obtained, first, because, owing to surface segregation, and, second, because the created surface layers have properties different from those of the bulk material. The well dispersed additives on the surface of the SnO_2 powders (see Figure 3) resist the mutual diffusion of SnO_2 necessary for the grain growth. We need to note that mentioned above effects are common for all metal oxides. The similar effect was observed for TiO_2 In_2O_3, WO_3, etc.[4,36]

DOPING INFLUENCE ON GAS SENSING CHARACTERISTICS

As it follows from the results presented in Figure 1 and Table 1, for attainment the desirable effect, for example the strong grain size decrease and the grain size stabilization during high temperature annealing, the concentration of those additives usually should be big enough. As a rule those concentrations exceed $5 - 10\%$.[29-33] At the same time in other works, where gas sensing properties of such doped materials were analyzed, it was established that commonly, the optimum of gas sensing properties is being observed at concentrations of doping additives less that $1 - 2\%$.[4,37-41] For example, Choi and Lee[42] for achievement better sensitivity to CH_4 used doping at the level ~ 0.1 wt.%. Results of other authors related to doping influence on gas sensing characteristics of the SnO_2-based sensors are shown in Figures 4 and 5.

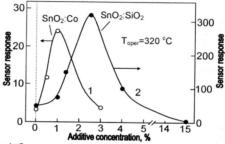

Figure 4. Additives influence on the response of doped SnO_2-based gas sensors: 1 - Responses of pure and Co-doped SnO_2 nanofibers to100 ppm H_2. Data from ref.[43]; 2 - Response of SnO_2:SiO_2-based sensors to 50 ppm EtOH as a function of the SiO_2 content. Data extracted from ref.[30]

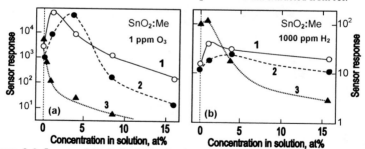

Figure 5. Influence of the SnO_2 doping by Co, Ni, Cu and Fe during film deposition by spray pyrolysis on sensor response to (a) ozone and (b) H_2: $T_{pyr} = 410–420$ °C; $d\sim45–55$ nm; 1: SnO_2:Fe; 2: SnO_2:Co; 3: SnO_2:Cu. Reprinted with permission from ref.[4]. Copyright 2005: Elsevier

These results indicate that introduction of high concentrations of additives leads to a sharp worsening of gas sensing properties in spite of strong decrease of the grain size.[30,43-47] This means that in doped metal oxides not only grain size controls gas sensing characteristics. There are other factors which can exert stronger influence on gas sensor parameters then the grain size. Besides, it was established that not all impurities, introduced in metal oxide matrix promote the inhibition of the grain's growth. Pavelko et al.,[48,49] for example, established that such bulk uncontrolled impurities as Cl$^-$, Na$^+$, SO$_4^-$ and Pd significantly accelerated the growth of the SnO$_2$ nanocrystallites (see Figure 6). Pavelko et al.[49] have also found that indicated impurities such as Cl and Na took part in the surface oxidation processes causing remarkable signal drift of sensor response. Acceleration of the SnO$_2$ grain growth at high temperature also takes place for doping by Cu, Co, and V.[36] The same effect was observed for other metal oxides including TiO$_2$.[36]

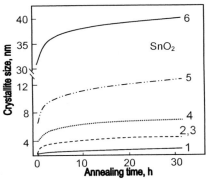

Figure 6. Dependence of the SnO$_2$ crystallite average size from the time of isothermal heating at 700°C: (1) SnO$_2$ nanopowder prepared by precipitation from tin acetate solution; (2) SnO$_2$ doped by S (0.036 wt%); (3) SnO$_2$ doped by Cl (0.038 wt%); (4) SnO$_2$ with surface modified by Pd (1.2 wt%); (5) SnO$_2$ doped by Pd (0.024 wt%); (6) commercial SnO$_2$ nanopowder from Sigma–Aldrich (specific surface area ~45 m^2/g). Data extracted from ref.[48]

It is important that during simultaneous study of gas sensing and catalytic properties of doped metal oxides it was established that as a rule the change of sensor response did not coincide with the change of catalytic activity of analyzed material (see Figure 7).[44] For undoped metal oxides as a rule the correlation between catalytic activity and sensor response is being observed. Such behavior of tested characteristics testifies that the observed decrease of sensor response of heavy doped metal oxides is not connected with reducing catalytical activity of sensing material, as it could be assumed at the first sight.

Korotcenkov and co-workers[50,51] analyzing influence of the bulk doping on luminescence properties of SnO$_2$ established that sensor sensitivity drop at superfluous concentration of doping additives was a result of the increase of the concentration of structural defects. It was concluded that the high level of the SnO$_2$ structure disordering, caused by doping, may be the reason of the increase of the concentration of the surface states, pinning the surface Fermi level and limiting the Fermi level shift during interaction with gas surrounding.[52,53] One should note that this conclusion was confirmed by theoretical[34] and experimental studies.[52] It was established that the growth of the concentration of doping additives in the SnO$_2$ was accompanied by the appearance of an additional spectral bands both in CL and in XRS spectra. According to Toimil Molares et al.,[54] the nature of those bands in XRS spectra is connected with structural disordering of the SnO$_2$ surface, which is responsible for the appearance of additional states inside a band gap. A considerable increase in half-width of the SnO$_2$ Raman peaks after Pt and Pd incorporation in the SnO$_2$ lattice ($C_{Pt,Pd}$ = 2 wt.%) observed by Toimil Molares et al.,[54] is another direct confirmation, that the SnO$_2$ lattice may be disturbed by the presence of additives at high concentrations.

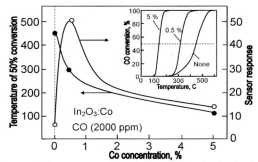

Figure 7. Influence of the concentration of Co additives in the In$_2$O$_3$ on the response and temperature of 50% CO conversion by SnO$_2$:Co-based sensor and catalyst. An aqueous solution of InCl$_3$ was hydrolyzed with ammonium hydroxide to produce In$_2$O$_3$, and the resulting precipitate was calcined at 850 °C. Metal oxides were added to the In$_2$O$_3$ powder by impregnating with an aqueous solution of each metal salt, followed by calcining at 600 °C. Catalytic oxidation of CO and H$_2$, 2000 ppm each with wet air, was carried out in a conventional fixed bed flow reactor. Data extracted from ref.[44]

Tian et al.[55] described the same processes for SnO$_2$ doped by Mn. They found that Mn in SnO$_2$ not only lowered the crystallite size but also degraded the crystallinity of the nanoparticles. As the Mn content increases, the intensity of XRD peaks decreases and FWHM increases, which indicates about degradation of the crystallinity. This means that Mn doping in SnO$_2$ produces crystal defects around the dopants and the charge imbalance arisen from this defect changes the stoichiometry of the materials. The distortion of the SnO$_2$ lattice was also observed by Aragón et al.[31] during doping by Cr. It was established that additional distortions was introduced by the substitution of Sn by Cr ions. In addition, it was found that the maximum distortion took place for the sample doped with 3 mol% Cr content. The 3 mol% Cr is assigned as the concentration where the regime change from the solubility to surface segregation of Cr ions occurs.[31]

Similar conclusions were made earlier while studying metal oxides aimed for design of varistors.[56] The data obtained have shown that doping the SnO$_2$ system with Cr$_2$O$_3$ at a level of 0.05% builds up an optimized barrier at the grain boundary. For Cr$_2$O$_3$ concentrations higher than 0.05%, the system loses its nonlinearity. It was concluded that the annihilation of the voltage barrier at the SnO$_2$ intergrain contacts was conditioned by the defect formation at the grain boundary. The deterioration of crystallinity of doped metal oxides was also observed by Ivanovskaya et al.[57]

In addition to surface Fermi level pinning, the surface states can also operate as intermediate centers for tunneling through intergrain potential barrier (see Figure 8). As it is known, tunnel component has weaker dependence on the height of potential barrier (Φ_b) in comparison with thermoelectronic emission. We need also to take into account that in heavy doped semiconductors with strong structural disorder the defect states in space charge region of potential barrier can operate as intermediate centers for tunneling similarly to surface states as well.

Thus, in the frame of discussed approach, as well as in case of sensors parameters optimization at the expense of the grain size decrease, we meet a contradiction, because the achievement of better thermal stability of structural properties of metal oxide ceramics could be accompanied by considerable worsening of their gas sensing properties. This means that using bulk doping for optimization of sensor parameters, very often one should choose either sensitivity, or stability. Moreover, we must expect that nanocomposites on the base of doped metal oxides with stable grain size would have properties unacceptable for gas sensor applications. For example, in case of SnO_2:Si at 15 wt% SiO_2 contents the SnO_2:SiO_2 nanocomposite becomes insulating.[30]

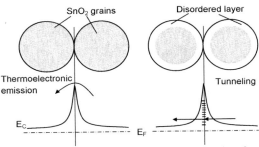

Figure 8. Diagram illustrating the influence of surface states at the interface on the mechanism of current transport.

Results obtained by Oswald et al.[58] also require our attention. Research carried out by Oswald et al.[58] has shown that samples of as-synthesized doped metal oxides are not in equilibrium state. For example, according to Oswald et al the thermodynamic equilibrium for doped SnO_2 single crystals was approached only after 24 h annealing above 850 °C and at 1000 °C for Sb and In, respectively. This means that the presence of doping additives, introduced in metal oxide for the grain size stabilization, by itself can be the reason of sensor parameters instability. The diffusion of doping elements during exploitation with following surface segregation can change significantly both electrophysical properties of the grains and conditions of the formation of intergrain contacts responsible for the height of intercrystallite's potential barrier. Therefore, the using of maximally low concentration of doping additives and long thermal treatment contributing to establishment of the thermodynamic equilibrium are important conditions required for achievement of high stability of sensors parameters designed on the base of doped metal oxides.

It is necessary to note that the demixing of doped SnO_2 during high temperature annealing, which is accompanied by the increase of the concentration of second phase, is common phenomena for metal oxides.[34] Usually it is stated that this effect takes place at T_{an}~800-900 °C. However, on the base of the results of the annealing influence on the lattice parameters of doped SnO_2 (see Figure 9) one can conclude that demixing takes place at lower temperature but with low rate. It is seen that the change of microstrains and the lattice parameters started already at T_{an}~500 °C.

Besides all said above we have also established the fact that mentioned above approach (grain size stabilizing through bulk doping of metal oxide), which was designed for both ceramic or thick film technologies, has limitation in some cases for the thin film technology.[47,60] It was shown that SnO_2 doping by Cu, Fe, Co during spray pyrolysis deposition did not promote the increase of thermal stability of the films' parameters. Due to increasing of the contents of the fine dispersed amorphous like phase, stronger structural changes during thermal treatments at temperatures 600-1000 °C take place in doped films in comparison with undoped ones. The mechanism of the appearance of the fine dispersed phase we have discussed earlier in refs.[47,61] According to suggested model the second oxide creates additional nucleation centers for the SnO$_2$ growth. Therefore, the growth of the SnO$_2$ film during deposition takes place not only due to the increasing in the size of crystallites incipient at primary stage of the growth, but also due to the appearance of the new grains, having considerably smaller size in comparison with already existing crystallites that appeared at the initial stages of the SnO$_2$ films' growth. This SnO$_2$ amorphous-like phase fills up the inter-crystallite space (see Figure 10) and promotes the densification of metal oxide matrix, i.e., the decrease of gas penetrability of deposited doped SnO$_2$ films. It is necessary to note that a possibility of the simultaneous presence in tin dioxide films of both crystallites and the fine dispersion amorphous-like phase was also experimentally proved by Jimenez et al.[62]

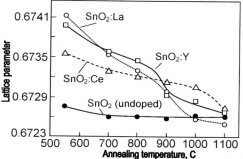

Figure 9. Lattice parameter and microstrain measurements as a function of the heat-treatment temperature. Powders of stannic oxide were prepared from precipitated hydrous β-stannic acid by oxidizing tin granules with nitric acid, evaporating the liquid and firing the powders at different temperatures under oxygen atmosphere. Data extracted from ref.[59]

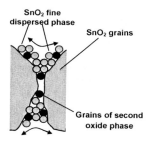

Figure 10. Schematic diagram of the inter-grain space containing fine dispersed phase. Adapted with permission from ref.[61]. Copyright 2009: Elsevier

More visually a disappearance of fine-dispersed phase after thermal treatment can be observed on AFM images of the SnO$_2$ films doped by copper (see Figure 11a,b). The diffuse image of grains' edges in metal oxide with big contents of fine dispersed phase appears due to low resolution of SEM technique, which does not allow discriminating the grains with small size located on the surface of basic oxide's crystallites. Data on the grain size changing during annealing process of undoped and

Cu-doped SnO$_2$ films are presented in Figure 11c. Those data were obtained using XRD measurements.

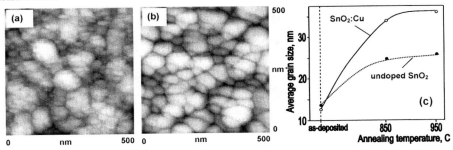

Figure 11. AFM images of doped SnO$_2$:Cu films deposited by spray pyrolysis before (a) and after (b) annealing at T_{an}= 850°C. (c) – Annealing influence on averaged grain size in undoped and Cu-doped SnO$_2$ films deposited by spray pyrolysis (T$_{pyr}$= 350°C; d~ 250 nm). Reprinted with permission from ref.[61]. Copyright 2009: Elsevier.

Because fine amorphous like phase fills intercrystallite space (see Figure 10), the coalescence of this fine phase with bigger crystallites increases the area of contact between intercrystallites, creating conditions at which conductive channel between crystallites is not being overlapped under any conditions. The last one should be accompanied by the decrease of sensor response. As it was shown in study conducted by Korotcenkov and Han,[61] the second oxide phase, presenting in the metal oxide matrix, did not hinder from this process, because crystallites' intergrowth took place through fine dispersed phase of based oxide. We need to note that Varela et al.[63] made a similar conclusion. They stated that very broad particle size distribution, i.e. the presence both big and small particles, first of all leads to the grain growth and agglomerates densifying.

CONCLUSIONS

Additional element in metal oxides can play an important role in the forming of their electrophysical, structural and gas sensing characteristics. However, while selecting both the type and the concentration of these additives, one should be very careful because we need to take into account many factors such as ability to form oxides of different oxidation states, catalytic activity, volatility, value and type of conductivity, and limited mutual solubility with based metal oxide. The last factor is especially important for structural engineering of gas sensing materials, since this parameter controls both the phase state of additives in based metal oxide, and structural disordering of the surface layer of the metal oxide grains. The concentration of additives in SnO$_2$ used for gas sensor design should not exceed 1-3 wt.%. At higher concentration, due to structural disorder, a strong decrease of sensor response takes place.

ACKNOWLEDGEMENTS

This work was supported by the Ministry of Education, Science and Technology of Korea (2011-0028736) and the Korea Science and Engineering Foundation (Grant No. R15-2008-006-01002-0).

REFERENCES

[1]C. Xu, J. Tamaki, N. Miura, and N. Yamazoe, Stabilization of SnO_2 Ultrafine Particles by Additives, *J. Mater. Sci.*, **27**(4), 963-971 (1992).

[2]G. Zhang and M. Liu, Effect of Particle Size and Dopant on Properties of SnO_2 –Based Gas Sensors, *Sens. Actuators B.*, **69**, 144–152 (2000).

[3]G. Korotcenkov, Metal Oxides for Solid-State gas Sensors: What Determines our Choice? *Mater. Sci. Eng. B.*, **139**, 1–23 (2007).

[4]G. Korotcenkov, Gas Response Control Through Structural and Chemical Modification of Metal Oxides: State of the Art and Approaches, *Sens. Actuators B.*, **107**(1), 209-32 (2005).

[5]V. Smatko, V. Golovanov, C.C. Liu, A. Kiv, D. Fuks, I. Donchev, and M. Ivanovskaya, Structural Stability of In_2O_3 Films as Sensor Materials, *J. Mater Sci.: Mater. Electron.*, **21**(4), 360-363 (2010).

[6]A.M. Taurino, M. Epifani, T. Toccoli, S. Iannotta, and P. Siciliano P. (2003) Innovative Aspects in Thin Film Technologies for Nanostructured Materials in Gas Sensor Devices, *Thin Solid Films*, **436**, 52–63 (2003).

[7]C. Xu, J. Tamaki, N. Miura, and N.Yamazoe, Grain Size Effects on Gas Sensitivity of Porous SnO_2-Based Elements, *Sens. Actuators B.*, **3**, 147-155 (1991).

[8]A. Gurlo, M. Ivanovskaya, N. Barsan, M. Schweizer-Berberich, U. Weimar, W. Gopel, and A. Dieguez, Grain Size Control in Nanocrystalline In_2O_3 Semiconductor Sensors, *Sens. Actuators B.*, **44**, 327-333 (1997).

[9]G. Korotcenkov, The Role of Morphology and Crystallographic Structure of Metal Oxides in Response of Conductometric-Type Gas Sensors, *Mater. Sci. Eng. R.*, **61**, 1-39 (2008).

[10]N. Barsan, M. Schweizer-Belberich, and W. Gopel, Fundamental and Practical Aspects in the Design of Nanoscaled SnO_2 Gas Sensors: A Status Report, *Fresenius' J. Anal. Chem.*, **365**, 287-304 (1999).

[11]A. Rothschild and Y. Komen, The Effect of Grain Size on the Sensitivity of Nanocrystalline Metal Oxide Gas Sensors, *J. Appl. Phys.*, **95**, 6374 (2004).

[12]M. Gillet, K. Aguir, M. Bendahan, and P. Mennini, Grain Size Effect in Sputtered Tungsten Trioxide Thin Films on the Sensitivity to Ozone, *Thin Solid Films*, **484**, 358-363 (2005).

[13]G. Korotcenkov, S.D. Han, B.K. Cho, and V. Brinzari, Grain Size Effects in Sensor Response of Nanostructured SnO_2- and In_2O_3-based Conductometric Gas Sensor, *Crit. Rev. Sol. St. Mater. Sci.*, **34**(1-2), 1-17 (2009).

[14]X. Wang, S.S. Yee, and W.P. Carey, Transition Between Neck-Controlled and Grain-Boundary-Controlled Sensitivity of Metal-Oxide Gas Sensors, *Sens. Actuators B.*, **24-25**, 454-457 (1995).

[15]N. Barsan and U. Weimar, Conduction Model of Metal Oxide Gas Sensors, *J. Electroceramics*, **7**, 143-167 (2001).

[16]D.E. Williams and K.F.E. Pratt, Microstructure Effects on the Response of Gas-Sensitive Resistors Based on Semiconducting Oxides, *Sens. Actuators B.*, **70**, 214-221 (2000).

[17]H. Ogawa, M. Nishikawa, and A. Abe: Hall Measurement Studies and an Electrical Conductive Model of Tin Oxide Ultrafine Particle Films, *J. Appl. Phys.*, **53**, 4448-4454 (1982).

[18]A. Gurlo, Interplay between O_2 and SnO_2: Oxygen Ionosorption and Spectroscopic Evidence for Adsorbed Oxygen, *ChemPhysChem.*, **7**, 2041-2052 (2006).

[19]V. Brynzari, G. Korotchenkov, and S. Dmitriev, Theoretical Study of Semiconductor Thin Film Gas Sensitivity: Attempt to Consistent Approach, *J. Electron. Technol.*, **33**, 225-235 (2000).

[20]N. Matsunaga, G. Sakai, K. Shimanoe, and N. Yamazoe, Formulation of Gas Diffusion Dynamics for Thin Film Semiconductor Gas Sensor Based on Simple Reaction–Diffusion Equation, *Sens. Actuators B.*, **96**, 226-233 (2003).

[21]N. Yamazoe and K. Shimanoe, Basic Approach to the Transducer Function of Oxide Semiconductor Gas Sensors, *Sens. Actuators B.*, **160**, 1352–1362 (2011).

[22]H. Yang, W. Jin, and L. Wang. Synthesis and Characterization of V_2O_5-doped SnO_2 Nanocrystallites for Oxygen-Sensing Properties, *Mater. Lett.*, **57**, 3686-3689 (2003).

[23] A.P. Maciel, P.N. Lisboa-Filho, E.R. Leite, C.O. Paiva-Santos, W.H. Schreiner, Y. Maniette, and E. Longo, Microstructural and Morphological Analysis of Pure and Ce-doped Tin Dioxide Nanoparticles. *J. Eur. Ceram. Soc.*, **23**, 707-713 (2003).

[24] B.-K. Min and S.-D. Choi, Role of CaO as Crystallite Growth Inhibitor in SnO$_2$, *Sens. Actuators B.*, **99**, 288-296 (2004).

[25] H. Yang, S. Han, L. Wang, I.J. Kim, and Y.M. Son, Preparation and Characterization of Indium-doped Tin Dioxide Nanocrystalline Powders, *Mater. Chem. Phys.*, **56**, 153-156 (1998).

[26] H. Miao, C. Ding, and H. Luo, Antimony-doped Tin Dioxide Nanometer Powders Prepared by the Hydrothermal Method, *Microelectron. Eng.*, **66**, 142-146 (2003).

[27] L.M. Fang, X.T. Zu, Z.J. Li, S. Zhu, C.M. Liu, W.L. Zhou, and L.M. Wang, Synthesis and Characteristics of Fe^{3+}-doped SnO$_2$ Nanoparticles *via* Sol–Gel-Calcination or Sol–Gel-Hydrothermal Route, *J. Alloys Comp.*, **454**, 261–267 (2008).

[28] L. M. Fang, X. T. Zu, Z. J. Li, S. Zhu, C. M. Liu, L. M. Wang, and F. Gao, Microstructure and Luminescence Properties of Co-doped SnO$_2$ Nanoparticles Synthesized by Hydrothermal Method, *J Mater. Sci: Mater. Electron.*, **19**, 868–874 (2008).

[29] I.T. Weber, R. Andrade, E.R. Leite, and E. Longo, A Study of the SnO$_2$·Nb$_2$O$_5$ System for an Ethanol Vapour Sensor: A Correlation Between Microstructure and Sensor Performance, *Sens. Actuators B.*, **72**, 180-183 (2001).

[30] A. Tricoli, M. Graf, and S.E. Pratsinis, Optimal Doping for Enhanced SnO$_2$ Sensitivity and Thermal Stability, *Adv. Funct. Mater.*, **18**, 1969–1976 (2008).

[31] F.H. Aragón, J.A.H. Coaquira, D.S. Candela, E. Baggio Saitovitch, P. Hidalgo, D. Gouvêa, and P.C. Morais, Structural and Hyperfine Properties of Cr-doped SnO$_2$ Nanoparticles, *J. Phys.: Conference Series,* **217**, 012079 (2010).

[32] A. Azam, A.S. Ahmed, M. Chaman, and A.H. Naqvi, Investigation of Electrical Properties of Mn Doped Tin Oxide Nanoparticles Using Impedance Spectroscopy, *J. Appl. Phys.*, **108**, 094329 (2010).

[33] N.L.V. Carreno, A.P. Maciel, E.R. Leite, P.N. Lisboa-Filho, E. Longo, A. Valentino, L.E.D. Probst, C.O. Paiva-Santos, and W.H. Schreiner, The Influence of Cations Segregation on the Methanol Decomposition on Nanostructured SnO$_2$, *Sens. Actuators B.*, **86**, 185-192 (2002).

[34] J. Nowotny, Surface Segregation of Defects in Oxide Ceramic Materials, *Solid State Ionics*, **28-30**, 1235-1243 (1988).

[35] P. Wynblatt, G.S. Rohrer, and F. Papillon, Grain Boundary Segregation in Oxide Ceramics, *J. Eur. Cer. Soc.*, **23**, 2841–2848 (2003).

[36] M. Ferroni, M.C. Carotta, V. Guidi, G. Martinelli, F. Ronconi., M. Sacerdoti, and E. Traversa, Preparation and Characterization of Nanosized Titania Fensing film, *Sens. Actuators B.*, **77**, 163-166 (2001).

[37] N. Yamazoe, New Approaches for Improving Semiconductor Gas Sensors, *Sens. Actuators B.*, **5**, 7-19, (1991).

[38] I. Matko, M. Gaidi, J.L. Hazemann, B. Chenevier, and M. Labeau, Electrical Properties under Polluting Gas (CO) of Pt- and Pd-doped Polycrystalline SnO$_2$ Thin Films: Analysis of the Netal Aggregate Size Effect, *Sens. Actuators B.*, **59**, 210-215 (1999).

[39] A. Dieguez, A. Vila, A. Cabot, A. Romano-Rodriguez, J.R. Morante, J. Kappler, N. Barsan, U. Weimar, and W. Gopel, Influence on the Gas Sensor Performances of the Metal Chemical States Introduced by Impregnation of Calcinated SnO$_2$ Sol-gel Nanocrystals, *Sens. Actuators B.*, **68**, 94-99 (2000).

[40] N.S. Ramgir, Y.K. Hwang, S.H. Jhung, I.S. Mulla, and J.S. Chang, Effect of Pt Concentration on the Physicochemical Properties and CO Sensing Activity of Mesostructured SnO$_2$, *Sens. Actuators B.*, **114**, 275-282 (2006).

[41]M. Yuasa, T. Masaki, T. Kida, K. Shimanoe, and N. Yamazoe, Nano-sized PdO Loaded SnO_2 Nanoparticles by Reverse Micelle Method for Highly Sensitive CO Gas Sensor, *Sens. Actuators B.,* **136**, 99-104 (2009).

[42]S.-D. Choi and D.-D. Lee, CH_4 Sensing Characteristics of K-, Ca-, Mg Impregnated SnO_2 Sensors, *Sens. Actuators B.,* **77**, 335-338 (2001).

[43]L. Liu, C. Guo, S. Li, L. Wang, Q. Dong, and W. Li. Improved H_2 Sensing Properties of Co-doped SnO_2 Nanofibers, *Sens. Actuators B.,* **150**, 806-810 (2010).

[44]H. Yamaura, K. Moriya, N. Miura, N. Yamazoe, Mechanism of Sensitivity Promotion in CO Sensor Using Indium Oxide and Cobalt Oxide, *Sens. Actuators B.,* **65**, 39-41 (2000).

[45]G. Korotcenkov, I. Boris, V. Brinzari, Yu. Luchkovsky, G. Karkotsky, V. Golovanov, A. Cornet, E. Rossinyol, J. Rodriguez, and A. Cirera, Gas Sensing Characteristics of One-Electrode Gas Sensors on the Base of Doped In_2O_3 Ceramics, *Sens. Actuators B.,* **103**, 13-22 (2004).

[46]G. Korotcenkov, I. Boris, A. Cornet, J. Rodriguez, A. Cirera, V. Golovanov, Yu. Lychkovsky, and G. Karkotsky, Influence of Additives on Gas Sensing and Structural Properties of In_2O_3- based Ceramics, *Sens. Actuators B.,* **120**, 657-664 (2007).

[47]G. Korotcenkov, V. Brinzari, and I. Boris, (Cu, Fe, Co or Ni)-doped SnO_2 Films Deposited by Spray Pyrolysis: Doping Influence on Film Morphology, *J. Mater. Sci.,* **43**(8), 2761-2770 (2008).

[48]R.G. Pavelko, A.A. Vasil'ev, V.G. Sevast'yanov, F. Gispert-Guirado, X. Vilanova, and N.T. Kuznetsov, Studies of Thermal Stability of Nanocrystalline SnO_2, ZrO_2 and SiC for Semiconductor and Thermocatalytic Gas Sensors, *Rus. J. Electrochem.,* **45**(4), 470–475 (2009).

[49]R.G. Pavelko, A.A. Vasiliev, E. Llobet, X. Vilanova, N. Barrabés, F. Medina, and V.G. Sevastyanov, Comparative Study of Nanocrystalline SnO_2 Materials for Gas Sensor Application: Thermal Stability and Catalytic Activity, *Sens. Actuators B.,* **137**, 637–643 (2009).

[50]G. Korotcenkov, M. Nazarov, M.V. Zamoryanskaya, M. Ivanov, A. Cirera, and K. Shimanoe, Cathodoluminescence Study of SnO_2 Powders Aimed for Gas Sensor Applications, *J. Mater. Sci. Eng. B.,* **130**(1-3), 200-205 (2006).

[51]G. Korotcenkov, B.K. Cho, M. Nazarov, D.-Y. Noh, and E. Kolesnikova, Cathodoluminescence Studies of Undoped and (Cu, Fe, and Co)-doped SnO_2 Films Deposited by Spray Pyrolysis Deposition, *Curr. Appl. Phys.,* **10**, 1123-1131 (2010).

[52]F. Hernandez-Ramirez, J.D. Prades, A. Tarancon, S. Barth, O. Casals, R. Jimenez–Diaz, E. Pellicer, J. Rodriguez, M.A. Juli, A. Romano-Rodriguez, J.R. Morante, S. Mathur, A. Helwig, J. Spannhake, and G. Mueller, Portable Microsensors Based on Individual SnO_2 Nanowires, *Nanotechnology* **18**, 495501 (2007).

[53]E. Comini, Metal Oxide Nano-Crystals for Gas Sensing, *Anal. Chim. Acta,* **568** (1–2), 28–40 (2006).

[54]M.E. Toimil Molares, A.G. Balogh, T.W. Cornelius, R. Neumann, and C. Trautmann, Fragmentation of Nanowires Driven by Rayleigh Instability, *Appl. Phys. Lett.,* **85**(22), 5337-5339 (2004).

[55]Z.M. Tian, S.L. Yuan, J.H. He, P. Li, S.Q. Zhang, C.H. Wang, Y.Q. Wang, S.Y. Yin, and L. Liu, Structure and Magnetic Properties in Mn Doped SnO_2 Nanoparticles Synthesized by Chemical Co-precipitation Method, *J. Alloys Compd.,* **466**, 26-30 (2008).

[56]P.R. Bueno, S.A. Pianaro, E.C. Pereira, L.O.S. Bulhoes, E. Longo, and J.A. Varela, Investigation of the Electrical Properties of SnO_2 Varistor System Using Impedance Spectroscopy, *J. Appl. Phys.,* **84**(7), 3700-3705 (1998).

[57]M.I. Ivanovskaya, D.A. Kotsikaua, A. Taurino, and P. Siciliano, Structural Distinctions of Fe_2O_3–In_2O_3 Composites Obtained by Various Sol–Gel Procedures, and Their Gas-Sensing Features, *Sens. Actuators B.,* **124**, 133–142 (2007).

[58]S. Oswald, G. Behr, D. Dobler, J. Werner, K. Wetzig, and W. Arabczyk, Specific Properties of Fine SnO_2 Powders Connected with Surface Segregation, *Anal. Bioanal. Chem.,* **378**, 411–415 (2004).

[59]R. Leite, A.P. Maciel, I.T. Weber, P.N. Lisbon-Filho, E. Longo, C.O. Paiva-Santos, A.V.C. Andrade, C.A. Pakoscimas, Y. Maniette, and W.H. Schreiner, Development of Metal Oxide Nanoparticles with

High Stability Against Particle Growth Using a Metastable Solid Solution, *Adv. Mater.*, **14**(12), 905-908 (2002).

[60]G. Korotcenkov and B.K. Cho, Thin Film SnO$_2$-based Gas Sensors: Film Thickness Influence. *Sens. Actuators B.*, **142**, 321-330 (2009).

[61]G. Korotcenkov and S.D. Han, (Cu, Fe, Co and Ni)-doped SnO$_2$ Films Deposited by Spray Pyrolysis: Doping Influence on Thermal Stability of SnO$_2$ Film Structure, *Mater. Chem. Phys.*, **113**, 756-763 (2009).

[62]V.M. Jimenez, J.P. Espinos, A. Caballero, L. Contreras, A. Fernandez, A. Justo, and A.R. Gonzalez-Elope, SnO$_2$ Thin Films Prepared by Ion Beam Induced CVD: Preparation and Characterization by X-ray Absorption Spectroscopy, *Thin Solid Films*, **353**, 113-123 (1999).

[63]J.A. Varela, O.J. Whittemore, and E. Longo, Pore Size Evolution During Sintering of Ceramic Oxides, *Ceram. Intern.*, **16**(3), 177-189 (1990).

EFFECT OF GLASS ADDITIVES ON THE DENSIFICATION AND MECHANICAL PROPERTIES OF HYDROXYAPAPTITE CERAMICS

Jiangfeng Song, Yong Liu*, Ying Zhang, and Zhi Lu
State Key Laboratory of Powder Metallugy, Central South University, Changsha 410083, China

ABSTRACT

Hydroxyapatite ceramic powders were sintered with glass additives in order to increase the sintering density and mechanical properties. Two types of Al_2O_3-Si_2O_3-B_2O_3 glass additives for hydroxyapatite (HA) have been tested, the combined effect of the sintering conditions and composition of glass, on the densification behavior, thermal stability, microstructure evolution and mechanical properties were studied. XRD, DSC and FESEM were performed to investigate the phase compositions as well as the microstructures of the as-sintered ceramics, the Vickers hardness and fracture toughness were also evaluated. The results show that the addition of glass induced the transformation of HA into β-TCP and α-TCP, relating to the chemical composition of the glass. Al_2O_3-Si_2O_3-B_2O_3 glass with a small amount of MgO (designated as MG) reveals a better densification and stability of β-TCP than the glass without MgO. HA ceramics with 5wt.% MG Al_2O_3-Si_2O_3-B_2O_3 glass sintered at 1300 $^{\circ}$C, show an optimistic mechanical properties, with a Vickers hardness of 5.26 ± 0.15 GPa, and a fracture toughness of 1.14 ± 0.19 MPa·m$^{1/2}$. Moreover, a resorbable β-TCP presented in the ceramics, show a potential applicant for biomedical fields.

INTRODUCTION

Calcium phosphate materials and hydroxyapatite ($Ca_{10}(PO_4)_6(OH)_2$, HA), in particular, have been widely investigated due to the excellent properties for biomedical applications. The properties include their similarity to the chemical composition of human bone and teeth, good bioactivity, biocompatibility, lack of inflammatory response as well as osteoconductive capacity[1]. However, HA has a problem of poor mechanical properties, such as brittleness and low fracture toughness, which can restrict its applications in non-load bearing or metallic implant surface coating[2].

Various approaches have been extensively explored to improve the mechanical properties of calcium phosphate ceramics[3]. Among the approaches, adding appropriate sintering additives is quite promising, especially for those additives which can induce the liquid phase sintering and enhanced densification[4]. Several glass sintering additives, such as CaO-P_2O_5 [5], SiO_2-CaO-P_2O_5 [1], CaO-P_2O_5-CaF_2 [6], SiO_2-B_2O_3 [7] and Al_2O_3-Si_2O_3-B_2O_3 [8], have already been studied for HA ceramics. In most cases, however, the introduced glass phase leads to a significant decomposition of HA with subsequent formation of tricalcium phosphate (β-TCP and α-TCP). Both β-TCP and α-TCP are known to biodegrade faster than HA, and as a consequence, the as-obtained calcium phosphate ceramics may provide several advantages over pure hydroxyapatite in some applications, namely, those requiring a high dissolution rate[6]. This decomposition process depends on the glass composition and the thermal treatment[1]. In this study, two kinds of Al_2O_3-Si_2O_3-B_2O_3 glass with different chemical composition are chosen as the sintering additives. The detailed chemical composition is refined, in order to generate a liquid phase at an appropriate temperature during the sintering procedure. It is difficult to sinter dense HA/β-TCP ceramics, for the phase transition temperature from β- to α-TCP is lower than a normal sintering temperature[9]. As a result, MgO and ZnO are chosen to stabilize the β-TCP, respectively. All the other components in the glass system are nontoxic because the resulting ceramics may have biomedical applications.

The mechanical properties of calcium phosphate ceramics highly depend on the phase composition and the characteristics of the original HA powder, including crystallinity, agglomeration, particle size, stoichiometry, and substitutions[10]. Based on our previous study[11], two kinds of HA powders with ultra different physical properties are chosen as the raw HA sources. According to the previous investigation, these pressureless sintered HA raw powders show no trace of dissociation to TCP even the sintering temperature is as high as 1280 °C, owing to their fine morphologies.

Since one of the most important limitations in the use of bioactive ceramics is their inadequate fracture toughness, the present work aimed at determining the hardness and fracture toughness of Al_2O_3-Si_2O_3-B_2O_3 glass-reinforced hydroxyapatite composites. The effect of glass composition and raw HA powders on the densification behavior, phase evolution and microstructure were investigated. Relationships between those mechanical properties and microstructural parameters, namely porosity and β and α tricalcium phosphate phases, were also established.

EXPERIMENTAL

Materials synthesis

HA powders were prepared according to a template method described elsewhere[11], PVP and CTAB were used as templates. Briefly, the calcium solution and phosphate solution were prepared by dissolving 1.0 mol $Ca(NO_3)_2 \cdot 4H_2O$ and 0.6 mol $NH_4H_2PO_4$ in 2000 ml distilled water, respectively. Then, PVP or CTAB was added into the calcium solution and the pH was adjusted to ~14. The phosphate solution was added dropwise to the above-mentioned solution under vigorously stirring at room temperature. The obtained suspension was aged and then filtrated, washed with distilled water and ethyl alcohol alternately several times. The white precipitate was subsequently dried in vacuum at 80 °C for 12 h to obtain HA powders. The obtained HA powders were marked as HA-PVP and HA-CTAB correspondingly.

The chemical compositions of the sintering additives were list in Table 1, Al_2O_3-Si_2O_3-B_2O_3 glasses contain ZnO and MgO were designated as ZG and MG, respectively. Analytical grade chemicals SiO_2, Al_2O_3, H_3BO_3, ZnO, Li_2CO_3, K_2CO_3, NaF and $Mg_2CO_3(OH)_2$ were used as the starting materials. The chosen starting materials were ball milled for 4 h to get a homogenized slurry, and the slurry was dried and sieved. Then the mixtures were melted in ZrO_2 crucibles at 1500 °C for 2 h, at a heating rate of 10 °C/min. The resulting melts were quenched in distilled water. Then the glass was ground in an agate mortar, ball milled for 3 h and sieved for further characterization.

Table 1 Chemical compositions of the additive glass (in wt.%)

Sample	Composition					
ZG	SiO_2	Al_2O_3	B_2O_3	ZnO	Li_2O	K_2O
	34	20	30	10	1	5
MG	SiO_2	Al_2O_3	B_2O_3	MgO	NaF	K_2O
	34	14	15	22	6	9

Sintering process

All the HA + 5 wt.% additive powder mixtures were prepared by mixing the HA powders with the selected additives thoroughly in a ball mill machine for 2 h. Table 2 lists the mixtures of different HA powders with different glass additives. For instance, HA powders templated by PVP with 5 wt.% addition of ZG were designated as HPZG.. For comparison, the HA powders without additives were also ball milled under the same condition. After dried and sieved, all the powders were biaxially compressed to pellets with a diameter of 12 mm in a stainless mold under 100 MPa. The compacts were sintered at 1200 °C ~ 1300 °C for 2 h, and the heating rate is 10 °C/min. All the samples were furnace cooled.

Table 2 Compositions of the bodies (in wt.%)

Sample	HA-PVP	ZG	MG	Sample	HA-CTAB	ZG	MG
HP	100	-	-	HC	100	-	-
HPZG	95	5	-	HCZG	95	5	-
HPMG	95	-	5	HCMG	95	-	5

Characterization

The synthesized glass additives were characterized by X-ray diffraction (XRD; D/ruax 2550PC). The mixed powders of HA and additives were characterized by thermalanalyzer (DSC; METTLER DSC/TGA 1). XRD was also performed to investigate the phase change and decomposition of as-sintered HA + 5 wt.% additive ceramics. Quantitative phase analysis was performed based on the adiabatic principle, using the following equation[12]:

$$W_X = \left(\frac{K_A^X}{I_X} \sum_{i=A}^{N} \frac{I_i}{K_A^i} \right) \tag{1}$$

where W_X is the weight fraction of X phase; A is the chosen arbitrary reference phase; K_A^X and K_A^i are the relative intensity of phase X and i as reference to phase A; I_X and I_i are the measured integrated intensities for phase X and i, respectively; N is the total amount of phases of the samples. K_A^X is equal to the RIR value ratio between X phase and A phase, the RIR values for all phases were excerpted from reference[13]. The microstructure was observed using the field emission scanning electron microscope (FESEM; NOVA NANOSEM 230). The samples were polished to a 1-μm finish, and then chemically etched for 3 min using 10 vol.% hydrochloric acid. The average grain size was determined from FESEM images using the quantitative metallographic method. The density was measured by Archimedes' method, the value was averaged of 3 valid samples for each group.

The Vickers hardness and indentation toughness were also characterized. The detailed measurements were given elsewhere[11]. The hardness and fracture toughness values were averaged by 5 valid measurements for each sample.

RESULTS

Phase composition

Figure 1 and 2 compare the XRD patterns of HA powders and HA powders with 5 wt.% sintering additives sintered at 1200 °C and 1300 °C, respectively. The HA powders sintered at both temperatures show no decomposition, all the peaks match the JCPDS pattern 09-0432. Depending upon the additives and the sintering temperature used, β-TCP (JCPDS 09-0169) and α-TCP (JCPDS 09-0348) were formed in the microstructure of the composites, due to the reaction between HA and the glass phase. The phase quantification results performed by the adiabatic method are presented in Table 3. For samples sintered with both kinds of additives, β-TCP was always present in the composites. α-TCP was only detected in samples with ZG Al_2O_3-Si_2O_3-B_2O_3 glass additives, especially for HA-PVP ones. The α-TCP was found in the HA-CTAB samples at the temperature as high as 1300 °C, indicate that HA-CTAB powders were less reactive than HA-PVP. The MG Al_2O_3-Si_2O_3-B_2O_3 glass additives caused a lower transformation of HA into TCP phases than ZG glass, showed to stabilize β-TCP.

Table 3 Phase composition of the ceramics (in wt.%)

| | 1200°C | | | 1250°C | | | 1300°C | | |
	HA	α-TCP	β-TCP	HA	α-TCP	β-TCP	HA	α-TCP	β-TCP
HP	100	0	0	100	0	0	100	0	0
HPZG	42.0	24.0	34.0	18.7	61.1	20.3	27.4	65.3	7.3
HPMG	58.6	0	41.4	49.5	0	50.5	66.5	0	33.5
HC	100	0	0	100	0	0	100	0	0
HCZG	47.3	0	52.7	6.7	0	93.3	45.3	31.2	23.5
HCMG	50.9	0	49.1	16.2	0	83.8	19.3	0	80.7

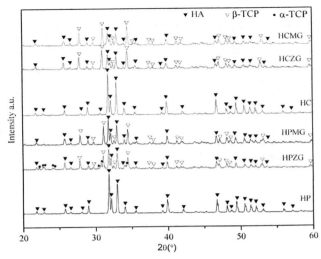

Fig. 1 XRD patterns of ceramics sintered from HA powders and HA powders with 5wt.% additives at 1200 °C

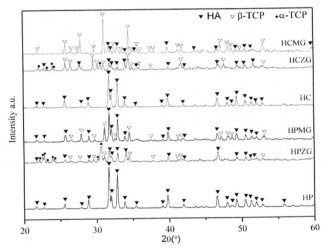

Fig. 2 XRD patterns of ceramics sintered from HA powders and HA powders with 5wt.% additives at 1300 °C

Densification

Fig. 3 corresponds to the thermal decomposition behaviors of the as-synthesized and the composite powders. The broad endothermic peaks appeared from 800 °C to 950 °C, were only found in ceramics with additives, not found in the pure HA ceramics. This indicates that the glass additives might react with the matrix HA at this temperature range. The peak may be attributed to the melting of the glass powders which result in a liquid phase leading to fast densification. Based on the glass transition temperatures (Tg) of the two glasses are about 540 °C[14] and 550 °C[15], the expecting melting temperature range of the glass powder approximately around 800 °C. Previous research[16; 17]revealed that the endothermic peak appearing at 800 °C was due to the increasing of HPO_4^{2-} group that was condensate to pyrophosphate (P_2O_7). This group reacted with the apatite finally to forming β-TCP, then transformed to α-TCP at high temperatures. This is consistent to the phase composition according to Fig. 1 and 2. In addition, the peak area of HA-PVP with additives is bigger than HA-CTAB, may be due to the fiercer reaction between HA and glass powders.

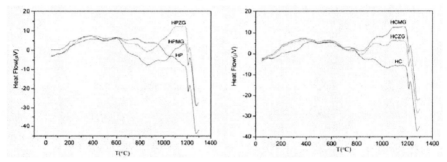

Fig.3 Thermal behavior of DSC of prepared powder samples

The relative density of all the ceramics sintered at 1200 °C, 1250 °C and 1300 °C are shown in Fig. 4. As the ceramics with additives decomposed to β-TCP and α-TCP, the relative density depends on the percentage of decomposition phases presented in each sample. The theoretical densities of HA , α-TCP[18] and β-TCP[19] are 3.156 g/cm³ , 2.868 g/cm³ and 3.049 g/cm³, respectively. All the densities increase with the sintering temperature, except for HCZG and HCMG, the mean density of HCZG and HCMG slightly decreases when the sintering temperature increase from 1250 °C to 1300 °C, although the decrease tends to overlap within the stand deviation of error. In addition, all the HA-PVP ceramics show relatively lower density than HA-CTAB ones at the same temperature, after the addition of sintering additives. Although the relative densities of HA-PVP and HA-CTAB sintered ceramics are comparable. This indicates that the chosen sintering additives cause the retard of the densification behavior for the HA-PVP powders. However, the additives can improve the densification of HA-CTAB powders. Consequently, the addition of MG or ZG is not good for all kinds of raw HA powders. The density of HPZG and HPMG sintered at 1300 °C is 2.40±0.02 g/cm³ and 2.80±0.09 g/cm³, the corresponding relative density is 81.10±0.52 and 90.08±2.73, respectively. It is concluded that the HA-PVP ceramics with the chosen additives is not fully densified, even the sintering temperature is as high as 1300 °C. This retard of densification is abnormal, as the glass additives usually accelerate the densification rate due to the liquid phase sintering theory[4; 20]. This could attributed to the reaction between HA-PVP with glass powders leading to the significant decomposition of HA, the resulting decomposition products is found to hinder the sintering process[21]. According to our previous detailed study[11], the pure HA-PVP show a lower sinterability than the pure HA-CTAB, which also could cause the retard of densification of the composite. This indicates that the sintering behavior highly depends on the properties of the raw HA powders. As for two kinds of glass additives, according to Fig. 4, MG Al_2O_3-Si_2O_3-B_2O_3 glass powders show optimistic influence on the densification in comparison with ZG powders.

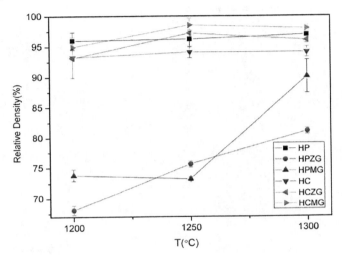

Fig. 4 Relative density as a function of sintering temperature

Microstructures

Figure 5 and 6 present the microstructures of ceramics sintered at 1200 °C and 1300 °C. According to the figures, the grain boundary of HA ceramics with sintering aids is not as clear as the pure HA, owing to the complicated phase composition of the composites. The grain sizes of all the samples measures by line intercept method are listed in Table 4. The grain sizes of HPZG are not available, because the densities of HPZG samples are relatively low. As the grain sizes are hard to measure, especially for the HA-CTAB with both additives, the BSED pictures of fracture surface is used to calculate the grain size. The typical BSED images for HCZG ceramics are shown in Fig. 7. It is found that grain sizes of all the samples increase with the sintering temperature according to Table 4, showing the same trend with the literatures[22; 23]. Interestingly, the grain size of HCMG samples has the lowest values of all the HA-CTAB samples when sintered at 1300 °C, which is averaged at 3.73 μm, 9% and 78% lower than HC and HCZG ceramics, respectively. However, the sintering aids show little influence on the grain growth, the ZG Al_2O_3-Si_2O_3-B_2O_3 glass additives even slightly accelerate the grain growth.

Table 4 Grain sizes of the sintered samples

	HP	HPZG	HPMG	HC	HCZG	HCMG
1200°C	1.81±0.37	--	3.01±0.86	2.04±0.37	1.37±0.30	1.94±0.75
1250°C	1.82±0.27	--	2.39±0.55	2.67±0.37	3.95±1.47	3.08±0.68
1300°C	3.07±0.41	--	5.37±1.78	4.07±0.27	6.65±1.03	3.73±0.73

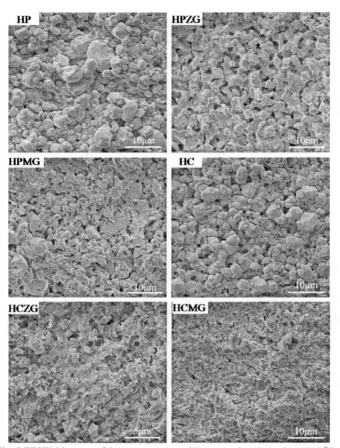

Fig. 5 FESEM images of the cross section of HA ceramics sintered at 1200 °C

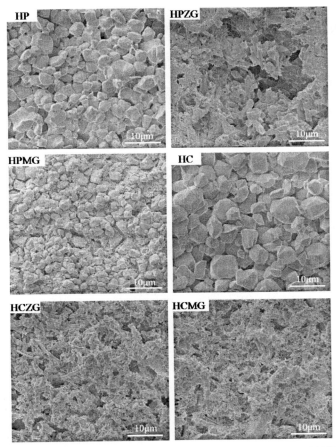

Fig. 6 FESEM images of the cross section of HA ceramics sintered at 1300 °C

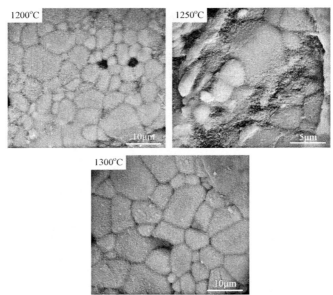

Fig. 7 Typical BSED FESEM images of the fracture surface of HCZG ceramics

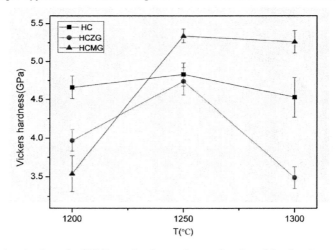

Fig. 8 Vickers hardness for CTAB templated ceramics as a function of the sintering temperature

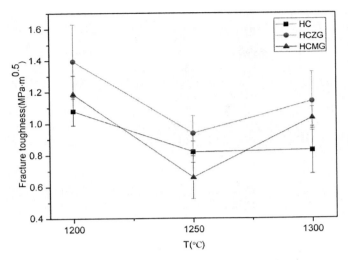

Fig. 9 Fracture toughness CTAB templated ceramics as a function of the sintering temperature

Mechanical properties

The mechanical properties of HA-PVP ceramics were not measured due to the relatively low density, especially for the HPZG sintered at 1200 °C and 1250 °C. The relationship of the Vickers hardness and the sintering temperature for HA-CTAB ceramics is shown in Fig. 8. The Vickers hardness of all the tested ceramics increase at first, and then decrease with the sintering temperature, which is due to the significant increase of grain size, showing the same trend in Ref.[23]. When the sintering temperature increases from 1250 °C to 1300 °C , the increase of grain size for HC , HCZG and HCMG are 52%, 68% and 21%, respectively, while the densities are in the range of 94.11-98.51%. HA-CTAB ceramics with MG Al_2O_3-Si_2O_3-B_2O_3 glass additives show an optimistic value of 5.34 ± 0.09 GPa and 5.26 ± 0.15 GPa at 1250 °C and 1300 °C, respectively.

The variation of the fracture toughness with the sintering temperature using the Evans' equation[24] can be seen in Fig. 9. It shows that the ceramics with sintering aids show a higher toughness value than ceramics without the aids, the only exception for this trend is HCMG ceramics sintered at 1250 °C. The higher mean toughness of ceramics with additives owing to the second phase in the composites. The second phase, β-TCP, is capable of enhancing the fracture toughness[25]. Because the phase transformation is accompanied by a volume increase, which creates residual stresses and finally result high fracture toughness. The only exception for this trend is HCMG ceramics sintered at 1250 °C, show a relatively low fracture toughness. This is may be due to the grain growth. Furthermore, the toughness of HCZG and HCMG decreases at first and then slightly increase with the sintering temperature. Based on previous reports, fracture toughness show a complicated relationship with the grain size [3; 11 26; 27]. According to Table 3, the amount of the secondary phase in HCZG and HCMG ceramics sintered at 1250 °C and 1300 °C is higher than that sintered at 1200 °C, leading to a higher toughness. This trend is maybe due to the combined effects of secondary phase and the grain size on the fracture toughness, otherwise, the mechanism of fracture behavior is still need to be improved.

DISCUSSION

Obviously, the Al_2O_3-Si_2O_3-B_2O_3 glass powders promote the decomposition of HA to TCP. Lopes et al.[25] and Wang et al.[8] reported that the addition of a phosphate-based glass and Al_2O_3-Si_2O_3-B_2O_3 glass could lead to the formation of TCP at high sintering temperatures, respectively. Thermal decomposition of hydroxyapatite is accomplished in two steps, dehydroxylation and decomposition[7], according to the following two equations, respectively:

$$Ca_{10}(PO_4)_6(OH)_2 \rightarrow Ca_{10}(PO_4)_6(OH)_{2-2x} + xH_2O_{gas} \tag{2}$$

$$Ca_{10}(PO_4)_6(OH)_2 \rightarrow 2Ca_3(PO_4)_2 + Ca_4P_2O_9 + H_2O_{gas} \tag{3}$$

The dehydroxylation (equation 2) of HA to oxyhydroxyapatite is a fully reversible reaction at temperatures above 800 °C. Decomposition to tricalcium phosphate and tetracalcium phosphate occurs at temperatures higher than 1000 °C (equation 3) after significant loss of hydroxyl groups. A study of stoichiometric hydroxyapatite revealed that the hydroxyapatite is stable after losing up to 75% of the hydroxyl ions in the form of water[28]. It is found from Fig. 3 that the HA composites present an endothermic peak at around 800 °C to 950 °C, while pure HA is stable. This reveals that the dehydroxylation of HA composites is more significant than pure HA, leading to the dissociation of HA to TCP.

It is evident from Fig.2, Fig.3 and Table 3 that the decomposition rate for all the composite samples increases when the temperature increases from 1200 °C to 1300 °C. This is consistent with the literatures[23; 29].

MgO is known as a sintering additive due to its ability to stabilize the β-TCP phase[9; 30]. Enderle et al[31] reported that the transformation temperature for magnesium-doped β-TCP ceramics increase with the content of Mg^{2+}. It is also reported that Zn^{2+} could stabilize the crystal lattice of β-TCP due to the substitution of Zina[32]. However, MgO seems to be more effective than ZnO on the stabilization of β-TCP, as trace magnesium is sufficient to nucleate the β-TCP phase[33]. According to the chemical composition of both glasses (Table 1), the major difference lies in the content of ZnO and MgO, 10 wt.% and 22wt.% for ZG and MG glass, respectively. Considering the lower content of ZnO than MgO, the MG glass powders show a better ability to stabilize β-TCP as a consequence. Furthermore, α-TCP is stable between 1125 °C and 1430 °C, a metastable phase at room temperature[31], which is often retransform to β-TCP when the samples were cooled. The detected α-TCP is due to the presence of SiO_2 in the glass powders, that is believed to be able to stabilize the α-TCP phase at room temperature[33]. The observation that HA-CTAB show a higher transformation temperature maybe due to its lower reactivity with the glassy phase than the HA-PVP. The reactivity of the ceramic powders is related to their physical properties. High reactivity is achieved with small particles which own a high specific surface area[34]. On the basis of our previous study[11], HA-PVP are irregular nanorods with an average length of ~100 nm and a mean diameter of ~20 nm, while HA-CTAB are regular prism-like with an average length of ~200 nm and a width of ~100 nm. Obviously, HA-PVP powders have a high specific surface area, leading to a high reactivity. It is concluded that the thermal stability is not only depend on the glass composition but also on the physical properties of the raw HA powders.

CONCLUSIONS

HA ceramics with 5 wt.% glass additives were successfully sintered. However, HA decompose to β-TCP and α-TCP due to the addition of the sintering aids. Additionally, MgO is able to stabilize the β-TCP phase during the sintering process. Sinterability as well as thermal stability of the ceramics highly depend on both the glass composition and the physical properties of the raw HA powders. The HA-CTAB ceramics with MG Al_2O_3-Si_2O_3-B_2O_3 glass powders additives(HCMG) sintered at 1300 °C, show an optimistic mechanical properties, with a Vickers hardness of 5.26 ± 0.15 GPa, and a fracture toughness of 1.14 ± 0.19 MPa·m$^{1/2}$. The ceramics contains a 80.7 wt.% of resorbable β-TCP, showing a potential applicants for biomedical fields.

ACKNOWLEDGEMENTS

The authors are grateful for the financial support from National Natural Science Foundation of China (NO. 50823006)

REFERENCES

[1]S. Padilla, J. Román, S. Sánchez-Salcedo, andM. Vallet-Regí, "Hydroxyapatite/SiO2-CaO-P2O5 glass materials: In vitro bioactivity and biocompatibility," *Acta Biomaterialia*, 2[3] 331-42 (2006).

[2]M. Mazaheri, M. Haghighatzadeh, A. M. Zahedi, andS. K. Sadrnezhaad, "Effect of a novel sintering process on mechanical properties of hydroxyapatite ceramics," *Journal of Alloys and Compounds*, 471[1-2] 180-84 (2009).

[3]J. Wang and L. L. Shaw, "Nanocrystalline hydroxyapatite with simultaneous enhancements in hardness and toughness," *Biomaterials*, 30[34] 6565-72 (2009).

[4]W. Suchanek, M. Yashima, M. Kakihana, andM. Yoshimura, "Hydroxyapatite ceramics with selected sintering additives," *Biomaterials*, 18[13] 923-33 (1997).

[5]M. A. Lopes, J. D. Santos, F. J. Monteiro, andJ. C. Knowles, "Glass-reinforced hydroxyapatite: A comprehensive study of the effect of glass composition on the crystallography of the composite," *Journal of Biomedical Materials Research*, 39[2] 244-51 (1998).

[6]A. C. Queiroz, J. D. Santos, F. J. Monteiro, andM. H. P. da Silva, "Dissolution studies of hydroxyapatite and glass-reinforced hydroxyapatite ceramics," *Mater Charact*, 50[2-3] 197-202 (2003).

[7]Y. Hu and X. Miao, "Comparison of hydroxyapatite ceramics and hydroxyapatite/borosilicate glass composites prepared by slip casting," *Ceram Int*, 30[7] 1787-91 (2004).

[8]Z. Wang, X. Chen, Y. Cai, andB. Lu, "[Influences of R2O-Al2O3-B2O3-SiO2 system glass and superfine alpha-Al2O3 on the sintering and phase transition of hydroxyapatite ceramics]," *Sheng Wu Yi Xue Gong Cheng Xue Za Zhi*, 20[2] 205-8 (2003).

[9]H.-S. Ryu, K. S. Hong, J.-K. Lee, D. J. Kim, J. H. Lee, B.-S. Chang, D.-h. Lee, C.-K. Lee, andS.-S. Chung, "Magnesia-doped HA/[beta]-TCP ceramics and evaluation of their biocompatibility," *Biomaterials*, 25[3] 393-401 (2004).

[10]D. Veljovic, B. Jokic, R. Petrovic, E. Palcevskis, A. Dindune, I. N. Mihailescu, andD. Janackovic, "Processing of dense nanostructured HAP ceramics by sintering and hot pressing," *Ceramics International*, 35[4] 1407-13 (2009).

[11]J. Song, Y. Liu, Y. Zhang, andL. Jiao, "Mechanical properties of hydroxyapatite ceramics sintered from powders with different morphologies," *Materials Science and Engineering: A,* 528[16-17] 5421-27 (2011).

[12]F. H. Chung, "Quantitative interpretation of X-ray diffraction patterns of mixtures. III. Simultaneous determination of a set of reference intensities," *Journal of Applied Crystallography,* 8[1] 17-19 (1975).

[13]P. Prevéy, "X-ray diffraction characterization of crystallinity and phase composition in plasma-sprayed hydroxyapatite coatings," *Journal of Thermal Spray Technology,* 9[3] 369-76 (2000).

[14]Z. X. Hou, S. H. Wang, Z. L. Xue, H. R. Lu, C. L. Niu, H. Wang, B. Sun, andC. H. Su, "Crystallization and microstructural characterization of B(2)O(3)-Al(2)O(3)-SiO(2) glass," *J Non-Cryst Solids,* 356[4-5] 201-07 (2010).

[15]E. Taheri-Nassaj, A. Faeghi-Nia, andV. K. Marghussian, "Effect of B2O3 on crystallization behavior and microstructure of MgO-SiO2-Al2O3-K2O-F glass-ceramics," *Ceram Int,* 33[5] 773-78 (2007).

[16]T. I. Ivanova, O. V. Frank-Kamenetskaya, A. B. Kol'tsov, andV. L. Ugolkov, "Crystal structure of calcium-deficient carbonated hydroxyapatite. Thermal decomposition," *Journal of Solid State Chemistry,* 160[2] 340-49 (2001).

[17]A. A. Mostafa, H. Oudadesse, M. B. Mohamed, E. S. Foad, Y. Le Gal, andG. Cathelineau, "Convenient approach of nanohydroxyapatite polymeric matrix composites," *Chemical Engineering Journal,* 153[1-3] 187-92 (2009).

[18]M. Yashima, Y. Kawaike, andM. Tanaka, "Determination of Precise Unit-Cell Parameters of the α-Tricalcium Phosphate Ca3(PO4)2 Through High-Resolution Synchrotron Powder Diffraction," *Journal of the American Ceramic Society,* 90[1] 272-74 (2007).

[19]M. Yashima, A. Sakai, T. Kamiyama, andA. Hoshikawa, "Crystal structure analysis of [beta]-tricalcium phosphate Ca3(PO4)2 by neutron powder diffraction," *Journal of Solid State Chemistry,* 175[2] 272-77 (2003).

[20]J. Wang and L. L.Shaw, "Morphology-Enhanced Low-Temperature Sintering of Nanocrystalline Hydroxyapatite," *Advanced Materials* 19 2364-69 (2007).

[21]Q. Chang, D. L. Chen, H. Q. Ru, X. Y. Yue, L. Yu, andC. P. Zhang, "Toughening mechanisms in iron-containing hydroxyapatite/titanium composites," *Biomaterials,* 31[7] 1493-501 (2010).

[22]S. Ramesh, C. Y. Tan, S. B. Bhaduri, W. D. Teng, andI. Sopyan, "Densification behaviour of nanocrystalline hydroxyapatite bioceramics," *Journal of Materials Processing Technology,* 206[1-3] 221-30 (2008).

[23]S. Nath, K. Biswas, K. Wang, R. K. Bordia, andB. Basu, "Sintering, Phase Stability, and Properties of Calcium Phosphate-Mullite Composites," *Journal of the American Ceramic Society,* 93[6] 1639-49 (2010).

[24]G.R.Anstis, P.Chantikul, B.R.Lawn, andD.B.Marshall, "A Critical Evaluation of Indentation Techniques for Measuring Fracture Toughness:I.Direct Crack Measurements," *journal of The American Ceramic Society,* 64[9] 533-38 (1981).

[25]M. A. Lopes, F. J. Monteiro, andJ. D. Santos, "Glass-reinforced hydroxyapatite composites: fracture toughness and hardness dependence on microstructural characteristics," *Biomaterials,* 20[21] 2085-90 (1999).

[26]K. A. Khalil, S. W. Kim, andH. Y. Kim, "Consolidation and mechanical properties of nanostructured hydroxyapatite-(ZrO2 + 3 mol% Y2O3) bioceramics by high-frequency induction heat sintering," *Materials Science and Engineering: A,* 456[1-2] 368-72 (2007).

[27]V. R. and S. D., "Processing and properties of nanophase non-oxide ceramics," *Materials Science and Engineering A,* 301[1] 59-68 (2001).

[28]T. P. Hoepfner, "The influence of the microstructure of sintered hydroxyapatite on the properties of hardness, fracture toughness, thermal expansion and the dielectric permittivity," *Michigan State University* (2001).

[29]C. Yang, Y.-k. Guo, andM.-l. Zhang, "Thermal decomposition and mechanical properties of hydroxyapatite ceramic," *Transactions of Nonferrous Metals Society of China,* 20[2] 254-58 (2010).

[30]A. C. S. Dantas, P. Greil, andF. A. Müller, "Effect of CO32− Incorporation on the Mechanical Properties of Wet Chemically Synthesized β-Tricalcium Phosphate (TCP) Ceramics," *Journal of the American Ceramic Society,* 91[3] 1030-33 (2008).

[31]R. Enderle, F. Götz-Neunhoeffer, M. Göbbels, F. A. Müller, andP. Greil, "Influence of magnesium doping on the phase transformation temperature of [beta]-TCP ceramics examined by Rietveld refinement," *Biomaterials,* 26[17] 3379-84 (2005).

[32]E. Boanini, M. Gazzano, andA. Bigi, "Ionic substitutions in calcium phosphates synthesized at low temperature," *Acta Biomaterialia,* 6[6] 1882-94 (2010).

[33]J. W. Reid, K. Fargo, J. A. Hendry, andM. Sayer, "The influence of trace magnesium content on the phase composition of silicon-stabilized calcium phosphate powders," *Materials Letters,* 61[18] 3851-54 (2007).

[34]J. Li, H. Liao, andL. Hermansson, "Sintering of partially-stabilized zirconia and partially-stabilized zirconia--hydroxyapatite composites by hot isostatic pressing and pressureless sintering," *Biomaterials,* 17[18] 1787-90 (1996).

Unconventional
Sintering Processes

FIELD ASSISTED SINTERING OF NANOMETRIC CERAMIC MATERIALS

U. Anselmi-Tamburini, F. Maglia, and I. Tredici
Department of Chemistry
University of Pavia, Italy

ABSTRACT
      Field Assisted Sintering (FAST) and Spark Plasma Sintering (SPS) have been emerging in the last few years as the techniques of choice for the synthesis of bulk ceramic nanocrystalline materials. Bulk ionic materials with extremely small grain size are producing great scientific and technological interest. The nanostructure, in fact, can alter drastically the physical properties and the phase equilibria in this class of materials, offering a new tool for controlling their functional properties. In this review the basic aspects of "nanoionics" are introduced, together with the main challenges that faces the production of bulk, fully dense ceramic materials with extremely small grain size. Some examples of bulk nanoionic materials obtained using the FAST-SPS approach are presented. Possible mechanisms that might enhance the densification of ionic materials in FAST are presented and discussed.
      The synthesis of ionic materials in nanocrystalline form experienced a rapid growth of interest in the last two decades. The reason is mostly technological: several ionic materials present in fact relevant functional properties that find use in countless technological applications[1]. The functional properties of this class of materials have been traditionally controlled and modified through an accurate control of point defects concentration and distribution[2]. However, the advancements in nanoscience have evidenced as the nanostructure may represent an alternative route for producing even larger modification in the basic physical properties[3]. This brought some authors to introduce the term "nanoionics"[4-6], suggesting the possibility that the nanostructure can produce a whole new family of functional materials, in analogy with what nanoelectronics did in the case of semiconducting materials.

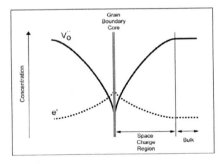

Figure 1. Example of defects distribution in a mixed conductor in the space charge region.

      In ionic materials the effects of nanostructure are mostly related to the interaction between bulk point defects and the surface (in powders) or the grain boundaries (in bulk material). In these materials, in fact, the external surfaces or the grain boundaries core bear always a net charge; this charge interacts with the charged defects producing a modification in their distribution (see Fig.1). The area where this interaction is observed is generally indicated by the term Space Charge Region (SCR). The extent of the SCR is defined by the following relation[7]:

$$\lambda^* = \lambda \left(\frac{4e\Delta\varphi}{kT} \right)^{1/2}$$

where λ indicates the Debye length:

$$\lambda = \left(\frac{\varepsilon\varepsilon_0 kT}{2e^2 C} \right)^{1/2}$$

ε is the bulk dielectric constant, ε_0 the vacuum permittivity, $\Delta\varphi$ the grain boundary core potential, and C the charge carrier concentration. When the grain boundary density becomes relevant, due to a reduction in grain size, the influence of the SCR on the overall defects equilibrium and distribution becomes significant, producing evident modifications in the physical properties of the material. Maier[8] divided the effects introduced by the nanostructure in ionic materials in two groups, that he identified by the terms "trivial effects" and "real effects". In the first group fall all the modifications resulting from the enhancement of some characteristics already present in the material, but that become more evident as a result of the increased amount of interfaces introduced by the nanostructure. "Real" effects, on the other hand, are associated with the appearance of new physical properties produced by the nanostructure. Several examples of "trivial" and "real" effects in ionic materials have been reported in the literature[4-8]. A typical example of "trivial" effect is represented by the localization of free electrons in the SCR of mixed conductors, such as ceria or titania. For extremely small grain size, the bulk conductivity in these materials is replaced by a conductivity confined along the grain boundaries region[9]. "Real" effects are expected to be observed when the grain size becomes comparable with the dimension of the SCR[10]. When this happens the regular bulk material disappears, producing a situation where the surface or interfacial properties prevail.

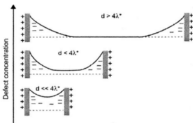

Figure 2. Size effect in defining the SCR morphology[4].

The influence of the nanostructure on the physical properties of ionic materials has been largely investigated on loose powders and on thin films, but very limited information is available on bulk materials. The reason lies in the difficulty of obtaining polycrystalline bulk materials characterized by a grain size small enough to produce significant modification of their physical properties. In ionic materials the SCR extends generally for only few nm, so that, in order to produce significant effects, a grain size between 10 and 30 nm is usually required[7]. The synthesis of bulk, fully dense materials with a grain size in this range represents a big challenge for the material scientists. The only viable approach is represented by the densification of nanopowders. Other approaches, such as the controlled crystallization of amorphous precursors, have found only very limited applications[11]. The densification of nanopowders of ceramic materials, however, is usually associated with a significant grain growth. Traditional sintering techniques, such as pressureless sintering and hot-pressing, involve long annealing times at high temperatures (usually around 2/3 of the melting point), since they are mostly driven by bulk diffusion. When these approaches are applied to the densification of ceramic nanopowders, grain sizes well above the limit required for the appearance of modification of the physical properties are usually obtained. A complicating element is represented by the quality of the nanopowders used for these applications. In the last two decades a large number of techniques for the production of ceramic nanopowders has been developed[12,13], including direct precipitation from aqueous solution, sol-gel methods, hydrothermal methods, polymeric precursors (such as Pechini) methods and propellant chemistry. These methods became immediately very popular since they offered a simple and affordable alternative to the earlier approaches, usually based on gas phase deposition[14]. Most of them allow to produce nanopowders with grain size well below 10 nm, but only rarely these are high quality monodispersed nanopowders as, almost invariably, they are characterized by high levels of agglomeration, that make the following densification process much more difficult.

Recently, however, non-conventional sintering approaches contributed significantly to improve the situation. Field Assisted Sintering techniques (FAST), such as Spark Plasma Sintering (SPS), have shown to be particularly effective in densifying the nanopowders of ceramic ionic materials[15-17]. These techniques share a combination of unique characteristics that help to reduce the grain growth during the densification process: in particular, they are generally characterized by processing times and temperatures that are much lower than in traditional sintering techniques. The terms Field Assisted Sintering refer to the possible role that electric fields or intense electric current might play in these processes; the details of such influence are still controversial and we will return on this point in the last part of this review.

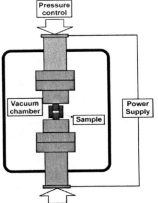

Figure 3. Schematic of a typical Field Assisted Sintering (FAST) apparatus.

A typical FAST apparatus follows the schematic reported in Fig. 3. The general design shows some similarities with the traditional hot-press: the sample, in powder form, is contained in a graphite die that is located between two hydraulic rams placed in a vacuum chamber, the main difference from hot-press is represented by the absence of furnaces or heating elements. The sample, in fact, is heated by a low voltage high intensity current flowing directly though the graphite die, that acts as the only heating element. This design allows high heating rates due to the low thermal inertia and the optimal thermal transfer; heating rates as high as 1000°C/min have been reported, with typical values being between 100 and 300°C/min. In commercial apparatuses such high heating rates are obtained using a low voltage (between 0 and 10 V) but high intensity currents, whose intensity can reach several thousand Amps, even up to 10.000 A in large systems. In early apparatuses pulsed currents with pulse size in the range of few milliseconds were used, while more recent apparatuses allow a broader control on the current waveform, allowing to use DC, AC, or pulsed currents.

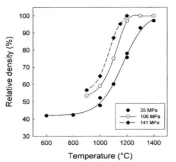 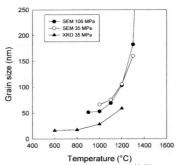

Figure 4. Example of FAST densification process on 8% Yttria Stabilized Zirconia[18-19].

An example of typical results obtained by the application of FAST on ceramic nanopowders is reported in Fig. 4[18-19]. In this example Tosoh 8% YSZ powders have been densified at different pressures and temperatures, but using the same sintering time of 5 min. It is remarkable that in such a short time full densification can be achieved using an applied pressure of just 100 MPa and a temperature below 1200°C. Pressureless sintering of the same material requires several hours at a temperature as high as 1400°C in order to achieve the same result. With such a short sintering time the grain growth results to be quite limited. As evidenced by Fig.4, at 1200°C the final grain size is just below 100 nm, a value only four times larger than the grain size of the starting powders (22 nm).

In order to achieve even smaller grain growth, however, these sintering conditions are not sufficient. It is evident from the data of Fig. 4, though, that an increase in the applied pressure results in a reduction of the temperature required to reach full densification. Since a reduction in temperature produces smaller grain size, it is expected that a significant reduction in the grain growth can be obtained applying much higher pressures. The compressive strength of graphite, however, does not generally allow pressures higher than 140 MPa. Materials characterized by higher compressive strength, such as Tungsten Carbide or, in alternative, double-stage dies similar to the one reported in Fig. 5, need to be used. With the assembly shown in Fig. 5 pressures as high as 1 GPa can easily be achieved[20]. The use of these high-pressure dies coupled with FAST machines is usually referred to as High-Pressure Field Assisted Sintering or HP-FAST. These double stage dies are composed by an external graphite element containing a smaller die whose plungers are made of silicon carbide or tungsten carbide. This assembly allows the production of samples with a diameter of 5 mm.

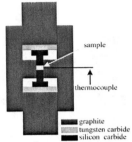

Figure 5. Schematic of a typical double-stage die used in High-Pressure Field Assisted Sintering (HP-FAST) apparatus[20].

The effect of high pressures on the grain growth of ceramic nanopowders is illustrated in Fig. 6.

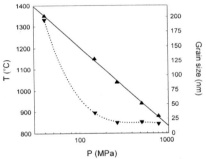

Figure 6. Dependence from pressure of the sintering temperature required to reach a relative density of 95%. Sintering time: 5 min. The grain size for each sample is also reported[20]. The grain size of the starting powder was 10 nm.

This figure reports the temperature required to obtain a relative density of 95% for applied pressures varying between 30 and 1000 MPa and an annealing time of 5 min. The results were obtained on an 8% YSZ powder obtained by the Pechini method and characterized by a grain size just above 10 nm. It is evident from this figure that an increase in pressure from 40 to 1000 MPa produces a decrease of more than 550°C in the temperature required to obtain the densification, resulting in a drastic reduction in the grain growth.

Figure 7. Typical nanostructure of a sample of 8% YSZ densified with HP-FAST. Sintering conditions: temperature 850°C, time 5 min, pressure 530 MPa. The relative density is > 98%.

Fig. 7 shows an example of a typical microstructure observed in fully densified nanocrystalline ionic materials produced by HP-FAST, while in Tab. 1 are listed the characteristics of the starting powders and of the densified materials, together with the densification conditions. It appears evident as fully dense materials characterized by a grain size between 10 and 17 nm can be obtained through this route. As shown in Fig. 7 these samples present a remarkably uniform nanostructure, despite the high level of agglomeration of the starting powders.

Table 1. Example of ionic oxides densified using HP-FAST[20].

Material	Powder grain size (nm)	Hold temperature. (°C)	Hold pressure (MPa)	Sintered sample grain size (nm)
CeO_2	7	625	600	11.6 ± 2.1
$Ce_{0.7}Sm_{0.3}O_2$	8	750	610	16.9 ± 3.1
YSZ (8%)	6.6	850	530	15.5 ± 3.2

The HP-FAST method offered for the first time the possibility to produce bulk fully dense ceramic materials with a grain size approaching 10 nm on a scale large enough to allow a systematic investigation of their physical properties. This triggered a flux of significant information on the role of nanostructure in defining the characteristics of bulk ionic materials. We will present here only a couple of examples both referring to zirconia.

The first example is relative to the influence of the nanostructure on the oxygen transport properties of bulk YSZ. YSZ is a well-known ionic electrolyte that finds widespread application in solid state electrochemical devices such as fuel cells or oxygen sensors. Because of these extensive technological applications there is much interest in any alteration of its transport properties, with particular attention to the possibility of reducing the grain boundary resistivity, that represents the main contribution to the overall resistivity in polycrystalline materials. This so called grain boundaries "blocking effect" derives, in high purity materials, mostly from the depletion of charge carriers in the SCR, originating from the repulsive electrostatic interaction between the positively charged grain boundary core and the charge carriers, represented by oxygen vacancies, $V_O^{\bullet\bullet}$ (see Fig. 1). The possible overlap between adjacent SCR, expected in nanocrystalline materials characterized by extremely small grain size, might in principle alter the potential barrier for the ion migration across the grain boundary and thus its blocking effects[8]. The results obtained so far, however, did not fully confirm these expectations[21]. Even in the case of grain size approaching 10 nm the specific grain boundary resistivity does show only limited modification. In this respect it has been confirmed the trend evidenced years ago by Guo and Zhang[22] - reported in Fig. 8 - that shows a gradual increase in the specific conductivity across the grain boundary as the grain size is decreased, with no evidence of discontinuities. There are two possible explanations for these results: the grain size achieved so far is not small enough to allow the appearance of "true" mesoscopic effects - Guo suggested a grain size approaching 5 nm would be necessary - or our understanding of the physics of the SCR is still too limited.

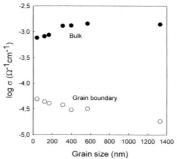

Figure 8. Dependence of bulk and specific grain boundary conductivity from the grain size for 3% YSZ at 550°C[22].

Although no significant modifications in the oxygen conductivity of YSZ have been observed, other relevant alterations of its transport properties as a result of the nanostructure have been recently reported. In particular, a new transport process has been evidenced in YSZ samples characterized by a grain size below 50 nm[23-25]. Samples characterized by this grain size exhibit a dramatic decrease in their resistivity when exposed to a moist environment at temperatures below 150°C. This low temperature enhanced conductivity appears to be ionic in nature and its magnitude is in direct relation with the grain size of the sample, becoming more evident for the smaller grain sizes and disappearing completely for grain size above 50 nm. The protonic nature of this conductivity was demonstrated by the observation of a relevant isotopic effect when deuterated water was used to humidify the carrier gas[23]. Fig. 9 shows the anomalous dependence of YSZ conductivity on temperature in a humid environment[24]. At high temperature, the typical YSZ oxygen ion conductivity is observed, but below 150°C a sudden inversion in the slope of the plot is observed, with an increment of conductivity that at room temperature results to be several orders of magnitude higher than that observed under dry atmosphere. It has also been shown that, at low temperatures, these materials are able to absorb a significant amount of water, up to 1 wt %. These results are quite surprising, as conventional micrometric stabilized zirconia is not a protonic conductor and does not absorb any water. The protonic solubility and mobility in bulk zirconia lattice is in fact very low[26], particularly at the low temperatures considered in this case. The close relationship existing between this new conduction mechanism and the grain size suggests an hydrogen incorporation and mobility that is most probably related with the grain boundary density. At the moment a clear understanding of the basic mechanism involved in this new process is not at hand and deeper investigations are still required.

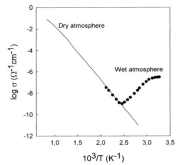

Figure 9. Dependence of conductivity on temperature for a sample of 8% YSZ with a grain size below 20 nm, when exposed to a flux of moist gas (P_{H_2O}=32 mbar) (symbols). The continuous line reproduces the temperature dependence of the conductivity for micrometric zirconia 8% YSZ[24].

Another striking example of modification of the physical properties induced by the nanostructure in ionic materials is represented by the alteration of phase equilibria in bulk zirconia. The stabilization of the high temperature cubic and tetragonal polymorph of zirconia is traditionally realized by alloying it with aliovalent oxides, such as CaO or rare earth oxides. In the last few years, however, it has been shown as the phase equilibria can be substantially modified by the nanostructure[27,28]. Stabilization of the tetragonal form has been reported in pure zirconia when the grain size was below some critical value[29]. Evidences of this phenomenon, however, have been initially shown only on loose powders and thin films, because these were the only forms of nanocrystalline zirconia that could be easily produced. Only recently it has been demonstrated that this stabilization could be achieved also in bulk materials. The possibility of stabilizing the tetragonal (or cubic) form in bulk, fully dense undoped zirconia, however, requires much more than the ability to maintain a small grains size in the densified material[30,31]. It is in fact the result of a complex balance between several contributions deriving primarily from the difference between the surface energy of the loose powders and the interfacial energy in the densified material and from the strain energy difference between the two forms, but it is influenced also by other factors, such as the external pressure, moisture, contamination and by the concentration of point defects. As a result, the process is very much dependent not only on the sintering conditions, but also on the characteristics of the starting powders.

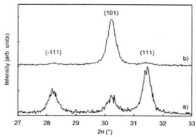

Figure 10. XRD relative to samples of fully densified undoped zirconia obtained using powders deriving from two different synthetic routes. a) powders deriving from hydrolysis of alkoxides; b) powders deriving from solvothermal synthesis in methanol. Both samples have been densified at 900°C under a pressure of 700 MPa[31]. The different stabilization of the crystallographic phases can be noticed from the relative peak intensities. (101): tetragonal phase; (-111), (111): monoclinic phase.

In a recent work Tredici et al.[31] have shown as powders obtained using different synthesis methods behave very differently in this respect. Synthetic methods involving aqueous solution, for instance, produce always very little level of stabilization of the tetragonal polymorph even in the case of very small grain size. On the other hand, the presence of agglomeration, usually considered detrimental in nanopowders, can play in this case a positive role. Through the appropriate combination of synthesis route and grain size, the authors have been able to produce fully dense bulk samples of undoped zirconia containing over 95% vol of tetragonal crystallographic form. This unprecedented result shows how the combination of an appropriate synthesis approach with FAST sintering can be used to produce bulk materials that retain characteristics typical of nanopowders and thin film. This opens the possibility to investigate the bulk properties of these materials, that have been inaccessible so far.

The two examples we briefly presented show how the FAST-SPS technique have been playing recently a relevant role in the technological and scientific investigation on the influence of nanostructure on the physical properties of bulk materials. Other examples can easily be found in the recent literature. The success of these applications, however, raises the questions of the basic foundations for such remarkable results.

The understanding of the physical basis of FAST has been object of much discussion in the FAST-SPS community in the last two decades[16]. Despite that, FAST and SPS still do not seem to be grounded on a firm scientific understanding. Many different mechanisms have been invoked to justify the results, but none of them seems to be able to explain all the aspects of the technique. The situation seems to get even more confusing with time. When it was first introduced, in fact, it was believed that most of the SPS results could be explained considering the formation of plasma, resulting from small sparks generated in the gaps between the particles by the high intensity pulsed current flowing through the sample[32] (Fig.11).

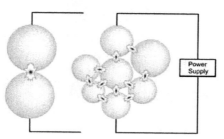

Figure 11. Possibile distribution of sparks between the sample particles during a FAST-SPS experiment.

In this model, localized plasma could enhance atomic mobility through vaporization and sputtering effects, resulting in cleaning of the particle surface and, ultimately, in an increased rate of neck growth, with effects extending also at the following stages of sintering. This interpretation enjoyed a large credit in the SPS community and for several years was considered the only viable explanation for the unusual characteristics of this techniques. However, it became soon clear that it suffered from a lack direct experimental evidences supporting it. The recent attempt by Hulbert et al.[33,34] to gather direct proof of electromagnetic plasma discharges during typical SPS processes, using a combination of different samples and experimental conditions, gave no positive results. Other authors suggested that localized thermal plasma could be generated at the point of contact between the particles by the intense Joule heating produced by the high intensity current flowing through the sample[35]. The possibility of formation of such localized and intense temperature gradients, however, has been generally proposed on the basis of oversimplified and unrealistic models for the heat generation and transfer[36]. Furthermore, these models for electromagnetic and thermal plasma generation cannot be invoked in the case of non-conducting samples, such as most ionic ceramic materials. This lack of experimental evidences supporting sparks and plasma leaves the field open for other possible mechanisms. Despite its relevance, this point received a surprisingly limited attention in the literature. Over the years, however, a number of possible phenomena have been proposed as alternative to the plasma effect. Most of them have been summarized in Tab. 2

Table 2: Mechanisms proposed for FAST sintering

Electromigration
Electric field induced diffusion
Temperature gradients
High heating rate
Pressure and stress gradient
Modification of defects concentration

An exhaustive discussion of all these mechanisms is far beyond the limits of this review. However, it must be noted that most likely the FAST-SPS process is far too complex to be explained on the basis of a single mechanism. This process is in fact much more complicated than conventional sintering. The most evident difference is represented by the presence of the electric fields and the electric currents, but the presence of temperature, strain and electric field gradients must also be considered. Furthermore, heating and cooling rates are much higher than in any other traditional solid-state processing technique, with the exclusion of microwave processing. As a result, a complex scenario involving different possible controlling mechanisms, depending on the experimental conditions and on the sample characteristics, must be considered.

The nature and the role of some of the mechanisms listed in Tab. 2 have been discussed in previous reviews[15-17] or have been analyzed in depth in the original literature[37, 38]. We would only like to note here that some of them are more general than others. Electromigration, for instance, might play a relevant role in the densification of metallic materials, but cannot be invoked in the case of ceramic materials; high heating rates and temperature gradients, on the opposite, might play a more general role. In general, however, very limited attention has been spent to the discussion of mechanisms that apply specifically to the case of ionic materials and to the peculiarity that this class of materials poses in this respect. FAST and SPS are characterized by the presence of quite strong electric fields and these fields can interact strongly with ionic materials. Most of them materials, for instance, show some level of ionic conductivity, at least at high temperature, resulting in what we indicated in Tab. 2 with the term *electric field induced diffusion*. This phenomena has been object of intense study in solid state electrochemistry[39,40], but has been poorly investigated in connection with sintering, so there is very little information about its role in enhancing the densification of ionic ceramic materials. In general terms, in presence of an external electric field, the flux of material can be described by the following equation[41]:

$$ J = -D\frac{\partial C_i}{\partial x} + Vq_i E $$

where D indicates the diffusion coefficient, C the concentration of the specie i, q the charge of the moving specie and E the electric field, acting as an external driving force. The effect of the electrotransport in ionic materials, however, is complicated by a number of considerations regarding not only the availability and the mobility of a charge carrier, but also the kinetics and the thermodynamics of the processes taking place at the electrodes[39,40]. As a result, the effect of an externally applied electric field depends not only on the mechanism of conduction but also on the nature of the electrodes. If the electrode is not able to accept or release the mobile species, no material flux can in fact be observed. Following the electrochemical terminology the electrodes can be divided into four groups: completely reversible (exchanging ions and electrons with the sample), semi-blocking ionic (exchanging only electrons), semi-blocking electronic (exchanging only ions) and totally blocking (no exchange of electrons and ions is allowed); this last case includes, obviously, the case of non-touching electrodes. In order to clarify the relative role of the various elements in the overall process and to identify possible connections with the sintering process, we will discuss a couple of examples referring to the schematic of Fig.12[42]

$$\frac{1}{a}M_aO_b \leftrightarrow \frac{b}{2a}O_2(g) + M^{|Z_1|(+)} + |Z_1|e'(Pt)$$

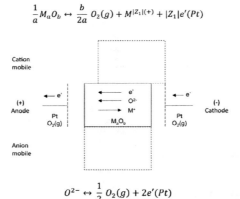

$$O^{2-} \leftrightarrow \frac{1}{2}O_2(g) + 2e'(Pt)$$

Figure 12. Schematic of possible transport mechanism resulting from the application of an electric field on an ionic sample in contact with two electrodes reversible for oxygen gas.

The figure summarizes the situation corresponding to two different materials in contact with electrodes allowing a reversible exchange of oxygen with the surrounding environment. In the lower part of the figure is described the situation corresponding to an oxide characterized by ionic conductivity through oxygen ions, as in YSZ. In this case, if a field is applied, oxygen ions are produced at the cathode, flow through the material (being the only mobile specie) and reach the other reversible electrode, that converts them back to oxygen gas. The overall result is a net transfer of oxygen from the cathodic to the anodic compartment. This process does not involve any annihilation or generation of crystallographic sites and it cannot produce any modification in the material microstructure. The upper part of the figure describes instead the case of an oxide where the only mobile specie is a cation, but the material is kept in contact with the same two electrodes described before, that allow only exchange of oxygen and not of the cation. The system can react to an imposed electric field decomposing crystallographic sites at the anode - releasing oxygen through the electrode - and producing cations that can travel through the sample (cations are the mobile species here) and reach the cathode where, acquiring oxygen from the electrode, can form a new crystallographic site. The result of this process is the decomposition of material at the anode and its reconstruction at the cathode; the sample, in this case, apparently moves towards the cathode. This macroscopic material transfer, involving annihilation and formation of crystallographic sites, can in principle produce a modification in the microstructure. It remains to be seen if this material transfer produced by the electrochemical process can be actually beneficial for sintering. Sintering, in fact, does involve a very specific transfer process, represented by the transport of material from the bulk of the grains towards the pores, in order to obtain densification. The material transport produced by the electric field, on the opposite, seems to transfer material from one electrode to the other, without any influence on the material in between. The situation, however, is more complex. There is, for sure, a growth of material at the cathode where that grows as a compact layer, without any porosity: the growth of single crystals through this route has in fact been demonstrated[42]. However, the flux of matter can produce a change in the microstructure also in other regions of the sample. In correspondence of pores and grain boundaries, in fact, a local non-zero flux divergence is produced ($\nabla J \neq 0$). As a result, local accumulation or depletion of material can be produced in analogy with what observed in the case of electromigration[43]. As an example, it has been reported by Byeon and Hong[42] that joining of single crystals is strongly enhanced in presence of an electric flow.

But electrotransport is not the only mechanism that can produce material transfer in ionic materials. Other more complex phenomena might come into play in the case of mixed conductors, when both ionic and electronic conductivities are possible. In these materials, in fact, the flux of a mobile ion J_i can be induced not only by the gradient of its own electrochemical potential $(\nabla \eta_i)$ but also by the electrochemical gradient of the electrons $(\nabla \eta_e)$. This relationship can be expressed using the following expression[45,46]:

$$\begin{pmatrix} J_i \\ J_e \end{pmatrix} = \begin{pmatrix} L_{ii} & L_{ie} \\ L_{ei} & L_{ee} \end{pmatrix} \begin{pmatrix} \nabla \eta_i \\ \nabla \eta_e \end{pmatrix}$$

where, following Onsager, $L_{ie} = L_{ei}$. In classical electrochemistry these cross-correlation terms are ignored, as each carrier is supposed to move independently from the others, but it has been shown recently[45] that for several semiconducting oxides these terms can be far from vanishing and in some cases L_{ie} are even bigger than L_{ii}. As a result, in presence of intense electronic fluxes a transport of cations is observed, and densification of porous pellets in proximity of one of the electrodes has been indeed reported in the case of CoO[45]. It is expected that also in this case the local non-zero flux divergence might play a role, producing modification of the microstructure not only in proximity of the electrodes, but also in any other region where a modification of the electronic flux is present, such as in the vicinity of pores and grain boundaries. The possible interaction between these transport mechanisms and the sintering process, however, has never been investigated properly, but it might play a significant role in FAST-SPS process, where high intensity current flows are always present.

Besides producing enhanced mobility of point defects, externally applied electric fields can also interact with larger defects in ionic materials, such as grain boundaries[46,47]. In ionic materials grain boundaries bear a net charge, as do the free surfaces, so under the influence of an external electric field these defects might enhance their mobility. Since the movement of grain boundaries is strictly associated with sintering, these phenomena might play a role in the possible interaction between applied field and sintering in ionic materials. Jeong et al. have investigated the details of this process in the case of alumina[46]. They evidenced as the effect of an externally applied field could be quite relevant and can be modeled considering an additional energetic term that sums to the capillary term in determining the movement of the grain boundary.

All the possible mechanisms here briefly described received little attention in relation with their role in sintering, so it is still premature to define how relevant is their influence on the densification of ceramic materials in FAST-SPS. More in-depth investigations are required on this regard. However, in discussing possible interactions between electric field and transport mechanisms, a critical point must be clarified first and it is represented by the actual values of field strength and current intensities experienced by the sample during the process. This is still a very controversial point due to the persisting relatively poor understanding of the basic aspects of these techniques. Field and current distribution in FAST-SPS depend, in fact, not only on the sample intrinsic electrical properties, but also - and to a large extent - on the sample microstructure, on the die geometry, and on the experimental conditions[48,49]. Even in the case of a fully dense conductive material, Anselmi-Tamburini et al.[48] have shown that the amount of current flowing through the material is a complex function of the geometrical and experimental parameters. In the typical setup of Fig. 3, sample and die represent two current paths in parallel to each other, so the main parameter controlling the amount of current flowing through the sample is the ratio between the resistance along these two paths. Samples characterized by a quite low electrical conductivity do not experience any significant current flow. However, geometrical factors must always be taken into account: the relative cross section between sample and die is of course relevant. Large samples processed in dies characterized by thin walls experience a higher current flux. Since the relative cross section of the die decreases when the sample diameter is increased, in order to avoid experimental assemblies that are too massive, larger samples experience more current flow. Even parameters apparently non-influent, such the sample thickness, the heating rate and the die thermal insulation, play a significant role in this respect. Further complications are introduced by the sample microstructure, that, evolving during the densification process, produces a change in current distributions throughout the process. It must finally be considered that, when complex die geometries are adopted, such as in the case of HP-FAST techniques[20], the possibility of current flux through the sample is very limited.

In the case of non-conducting samples no current is expected to flow through the sample; in this case the attention must be focused on the field strength. SPS-FAST apparatus typically applies only few volts across the entire assembly, so the macroscopic field strength experienced by the sample is actually quite small and is further reduced if we consider that the samples are contained in dies of an electrically conducting material. In contrast, some of the mechanisms described before require very intense electric field in order to produce significant effects in the short times usually involved in FAST processes. However, it must be noted that most non-conductive ionic materials are also good dielectric materials. Polarization effects in dielectric materials can produce a significant, yet localized, enhancement of an externally applied electric field. According to Techaumnat and Takuma[50] a chain of spherical dielectric particles can produce at the point of contact a field amplification of several orders of magnitude if the dielectric constant of the material is high enough (see Fig.13). This might lead to local field intensities in the range of values high enough to produce a score of effects, at least in the early stages of the process.

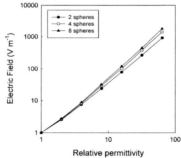

Figure 13. Electric field strength at the point of contact between a chain of 2, 4 and 8 spheres as a function of the relative dielectric constant of the material. The applied electric filed is 1 Vm^{-1}.[50]

In conclusion, the introduction of FAST techniques represented a significant advancement in the investigation on ionic nanomaterials, allowing a simple and reliable approach towards the synthesis of bulk, fully dense materials characterized by a grain size that, in some instances, approaches 10 nm. This remarkable achievement opens the possibility to investigate mechanical and electrical bulk properties of materials whose physical properties might be influenced by the nanostructure. The basic mechanisms involved in this densification method, however, are still debated. In the case of ionic materials the possible role of electric field and electric currents in enhancing the densification process still needs to be clarified.

REFERENCES
1. S.A. Wilson, R.P.J. Jourdain, Qi Zhang, R.A. Dorey, C.R. Bowen, M. Willander, Q.Ul Wahab, M. Willander, S.M. Al-hilli, O. Nur, E. Quandt, C. Johansson, E. Pagounis, M. Kohl, J. Matovic, B. Samel, W. van der Wijngaart, E.W.H. Jager, D. Carlsson, Z. Djinovic, M. Wegener, C. Moldovan, R. Iosub, E. Abad, M. Wendlandt, C. Rusu, K. Persson, "New materials for micro-scale sensors and actuators. An engineering review", *Mat.Sci.Eng. R,* 56 1–129 (2007).
2. Harry L. Tuller and Sean R. Bishop, "Point Defects in Oxides: Tailoring Materials Through Defect Engineering", Annu. Rev. Mater. Res. 41, 369-398 (2011).
3. H. Gleiter, "Nanostructured Materials: Basic Concepts and Microstructure", *Acta Mater.* 48, 1-29 (2000)
4. J. Maier, "Nanoionics: ion transport and electrochemical storage in confined systems", *Nature Materials* 4, 805 - 815 (2005)
5. J.Maier, "Defect chemistry and ion transport in nanostructured materials Part II. Aspects of nanoionics", *Solid State Ionics* 157, 327– 334 (2003)
6. J. Schoonman, "Nanoionics", *Solid State Ionics* 157, 319– 326 (2003)
7. X. Guo, R. Waser, "Electrical properties of the grain boundaries of oxygen ion conductors: Acceptor-doped zirconia and ceria", *Progress in Materials Science* 51, 151–210 (2006)
8. J. Maier, "Nanoionics: ionic charge carrier in small systems" *Phys.Chem.Chem.Phys.,* 11, 3011-3022 (2009).
9. S.Kim and J.Maier, "On the Conductivity Mechanism of Nanocrystalline Ceria", *J. Electrochem. Soc.,* 149, J73-J83 (2002).
10. P.Balaya, J.Jamnik, J.Fleig, and J. Maier, "Mesoscopic Hole Conduction in Nanocrystalline SrTiO$_3$. A Detailed Analysis by Impedance Spectroscopy", *Journal of The Electrochemical Society,* 154, P69-P76 (2007).
11. D.Li, H.Zhou, I.Honma, "Design and synthesis of self-ordered mesoporous nanocomposite through controlled *in-situ* crystallization", *Nature Materials* 3, 65 - 72 (2004).
12. D.Segal, Chemical synthesis of ceramic materials, *J. Mat. Chem.,* 7, 1297-1305 (1997)
13. G.Cao, "Nanostructures & Nanomaterials. Synthesis, Properties & Applications", Imperial College Press (2004).
14. H.Hahn, "Gas Phase Synthesis Of Nanocrystalline Materials", *Nanostructured Materials* 9, 3-12 (1997)
15. Z.A. Munir, U. Anselmi-Tamburini, M. Ohyanagi, "The effect of electric field and pressure on the synthesis and consolidation of materials: A review of the spark plasma sintering method", *J.Mater.Sci.* 41, 763–777 (2006)
16. R.Orrù, R.Licheri, A.M.Locci, A.Cincotti, G.Cao, "Consolidation/synthesis of materials by electric current activated/assisted sintering", *Mat.Sci.Eng. R*, 63, 127-287 (2009).
17. M. Omori, "Sintering, consolidation, reaction and crystal growth by the spark plasma system (SPS)", Mat.Sci.Eng. A, 287, 183-188 (2000)
18. U. Anselmi-Tamburini, J. E. Garay, Z. A. Munir, A.Tacca, F. Maglia, G. Spinolo, "Spark plasma sintering and characterization of bulk nanostructured fully stabilized zirconia: Part I. Densification studies", *J.Mat.Res.,* 19 (2004) 3255.
19. U. Anselmi-Tamburini, J. E. Garay, Z. A. Munir, A. Tacca, F. Maglia, G. Chiodelli, and G. Spinolo, "Spark plasma sintering and characterization of bulk nanostructured fully stabilized zirconia: Part II. Characterization studies", *J. Mater. Res.,* 19 (2004) 3263.
20. U. Anselmi-Tamburini, J.E. Garay, and Z.A. Munir, "Fast low-temperature consolidation of bulk nanometric ceramic materials", *Scripta Mater.,* 54, 823-828 (2006).
21. X. Guo, "Can we achieve significantly higher ionic conductivity in nanostructured zirconia?", *Scripta Mater.,* 65, 96-101 (2011)

22. X.Guo, Z.Zhang, "Grain size dependent grain boundary defect structure: case of doped zirconia", *Acta Mater.* 51, 2539–2547 (2003).
23. U.Anselmi-Tamburini, F.Maglia, G.Chiodelli, P.Riello, S.Bucella, Z.A.Munir, "Enhanced protonic conductivity in fully dense nanometric cubic zirconia", *Appl. Phys. Lett.*, 89, 163116 (2006)
24. G.Chiodelli, F.Maglia, U.Anselmi-Tamburini, Z.A.Munir, "Characterization of low temperature protonic conductivity in bulk nanocrystalline fully stabilized zirconia", *Solid State Ionics* 180, 297–301 (2009)
25. J.-S.Park, Y.-B.Kim, J.-H. Shim, S. Kang, T. M. Gur, and F.B. Prinz, "Evidence of Proton Transport in Atomic Layer Deposited Yttria-Stabilized Zirconia Films", *Chem. Mater.*, 22, 5366–5370 (2010).
26. Y. Nigara, K. Yashiro, J.-O. Hong, T. Kawada, J. Mizusaki, "Hydrogen permeability of YSZ single crystals at high temperatures", Solid State Ionics 171, 61–67 (2004).
27. H.Zhang and J.F. Banfield, "Thermodynamic analysis of phase stability of nanocrystalline titania", *J. Mater. Chem.*, 8, 2073-2076 (1998).
28. J. M. McHale, A. Auroux, A. J. Perrotta, A. Navrotsky, "Surface Energies and Thermodynamic Phase Stability in Nanocrystalline Aluminas", *Science*, 277, 788-791 (1997).
29. S.Shukla, S.Seal, "Mechanisms of room temperature metastable tetragonal phase stabilisation in zirconia", *Int Mater.Rev.*,50, 1–20 (2005).
30. F. Maglia, M. Dapiaggi, I.Tredici, B. Maroni, U.Anselmi-Tamburini, "Synthesis of fully dense nano-stabilized undoped tetragonal zirconia", J.Am.Cerm.Soc. 93, 2092-2097 (2010).
31. I.G.Tredici, F.Maglia, U.Anselmi-Tamburini, "Synthesis of bulk tetragonal zirconia without stabilizer: the role of the precursor nanopowders", J. Europ. Ceram. Soc., 32, 343–352 (2012).
32. M.Tokita, "Trends in Advanced SPS Spark Plasma Sintering Systems and Technology", *J.Soc.Powder Technol.Jpn.*, 30, 790 (1993)
33. D.M. Hulbert, A. Anders, D. V. Dudina, J. Andersson, D. Jiang, C. Unuvar, U. Anselmi-Tamburini, E. J. Lavernia, and A.K. Mukherjee, "The Absence of Plasma in "Spark Plasma Sintering", *J. Appl. Phys.* 104, 033305 (2008).
34. D.M. Hulbert, A. Anders, J. Andersson, E.J. Lavernia and A.K. Mukherjee, "A discussion on the absence of plasma in spark plasma sintering", *Scripta Mater.* 60, 835-838 (2009)
35. X.Song, X. Liu, J.Zhang, "Neck Formation and Self-Adjusting Mechanism of Neck Growth of Conducting Powders in Spark Plasma Sintering", *J. Am. Ceram. Soc.*, 89, 494–500 (2006).
36. T.B. Holland, U.Anselmi-Tamburini, D.V. Quach, T.B. Tran, and A.K. Mukherjee, "Electric Field Assisted Sintering. Part I: On the Local Field Strengths During Early Stage Sintering of Ionic Ceramics", in press
37. E.Olevsky and L.Froyen, "Constitutive modeling of spark-plasma sintering of conductive materials", *Scripta Mater.*, 55, 1175–1178 (2006)
38. E.Olevsky and L.Froyen, "Impact of Thermal Diffusion on Densification During SPS", *J. Am. Ceram. Soc.*, 92 S122–S132 (2009)
39. F. A. Kroger, "The Chemistry of Imperfect Crystals", North-Holland, Amsterdam; Interscience (Wiley), New York, 1964, p. 131.
40. H. Schmalzried: "Chemical kinetics of solids", VCH, Weinheim.
41. W. Jost, *Diffusion in Solids, Liquids, Gases*. Academic Press, 1952.
42. S.C. Byeon and K.S. Hong, "Electric field assisted bonding of ceramics," *Mat.Sci.Eng. A*, 287, 159-170 (2000).
43. P.S. Ho and T. Kwok, "Electromigration in metals," *Rep. Prog. Phys.*, 52, 301-348 (1989).
44. H.-I. Yoo, J.-H. Lee, M. Martin, J. Janek, e H. Schmalzried, "Experimental evidence of the interference between ionic and electronic flows in an oxide with prevailing electronic conduction", *Solid State Ionics*, 317-322, 67 (1994).

45. H.-I. Yoo e D.-K. Lee, "Onsager coefficients of mixed ionic electronic conduction in oxides", *Solid State Ionics*, 837-841, 179 (2008).
46. J.-W. Jeong, J.-H. Han, and D.-Y. Kim, "Effect of electric field on the migration of grain boundaries in alumina," *J.Am.Ceram.Soc.* 83, 915-918 (2000).
47. J.-I. Choi, J.-H. Han, and D.-Y. Kim, "Effect of titania and lithia doping on the boundary migration of alumina under an electric field," *J.Am.Ceram.Soc.*, 86, 640-643 (2003).
48. U. Anselmi-Tamburini, S.Gennari, J.E.Garay, and Z.A.Munir, "Fundamental Investigations on the Spark Plasma Sintering/Synthesis Process: II. Modeling of Current and Temperature Distributions", *Mat.Sci.Eng.A*, 394, 139-148 (2005).
49. K. Vanmeensel, A. Laptev, J. Hennicke, J. Vleugels, O. Van der Biest, "Modelling of the temperature distribution during field assisted sintering", *Acta Mater.*, 53, 4379–4388 (2005)
50. B. Techaumnat and T. Takuma, "Calculation of the electric field for lined-up spherical dielectric particles", *IEEE Trans.Dielec.Elec.Insul.*, 10, 623-633 (2003).

FABRICATION OF COPPER-GRAPHITE COMPOSITES BY SPARK PLASMA SINTERING AND ITS CHARACTERIZATION

Bunyod Allabergenov[a], Oybek Tursunkulov[a], Soo Jeong Jo[a], Amir Abidov[a], Christian Gomez[b], Sung Bum Park[a] and Sungjin Kim[a]

[a]School of Advanced Materials and System Engineering, Kumoh National Institute of Technology, 1 Yangho-dong, Gumi, 730-701, Korea.

[b]Facultad de Ciencias Quimicas, Universidad Autonoma de San Luis Potosi, Av. Manuel Nava #6, San Luis Potosi, S.L.P. 78290 Mexico.

ABSTRACT

The copper-graphite composite material with varying size of graphite powders and copper powders with optimal ratio were prepared by spark plasma sintering. The physical, mechanical, and electrical properties are directly depends to the size of the initial powders. In particular, at different sintering temperatures of SPS, and applying the optimal copper-graphite ratio (Cu-G) was fabricated samples from copper-derived composites where used of the graphite source and electrolytic grade copper powder. When particle size of the copper and graphite initial components was bigger than several microns, the mechanical and electrical properties of fabricated samples were deteriorated and lead to decrease qualitative characteristics. Thermal and composition parameters and the obtained composite samples fabricated by spark plasma sintering have been investigated by micro-hardness tester, hall measure system, density tester, optical microscope and FE-SEM.

1. INTRODUCTION

At present time with the technology developments of railways and maglev system, speedup of trains and cost reduction of maintenance facilities necessity to develop composite materials with excellent combination of mechanical and electrical properties are strongly required [1-7]. Among this composite materials copper-graphite are attractive materials for high-speed electric railways which has excellent combination of mechanical and electrical properties, in particular, an excellent sliding capability, great electric and heat conductibility, largely corrosion proof, high mechanical strength and good damping properties. These physical properties of copper-carbon composites have been exploited for many decades to transmit power from overhead contact wires and/or power rails to electrically operated vehicles. That is why nowadays these composites are widely used as brushes for motor technology, contact strips for pantographs and collector shoes in electric railways [8-12]. These composites are usually prepared by a technology based on powder metallurgy [13, 14]. This is because powder metallurgical processes offer the possibility of obtaining uniform brushes and of reducing the tedious and costly machining processes. However, this technology has certain limitations mainly related with the poor affinity between copper and graphite, which gives rise to weak interfaces with the consequent negative effect on the structural, mechanical and electrical properties of the material. That is why in this paper we suggested to use sintering processes for fabrication copper-graphite composite. The main aim of this research was to investigate the influence of processing parameters on the mechanical and electrical properties of copper–graphite composites made by using spark plasma sintering method. Mechanical and electric parameters and the obtained composite samples fabricated by SPS have been investigated by micro-hardness tester, hall system, density tester, optical microscope and FE-SEM.

151

2.   EXPERIMENTAL PROCEDURE

2.1. Materials

It was carried out two set of experiments for obtaining copper-graphite composite materials with comparatively high mechanical and electrical properties. Different type and varied particle size of carbon materials and copper powder were used for each set of the experiment. In particular, for the first set of experiments as initial materials were used spherical copper powder (CUSP50) which with average particle size 4.8-5μm and density of 8.9000 g/cm^3 and carbon and graphite powder (KB-3) with average size 30-40nm with 99.9% purity. The graphite powder (SAMCHON Pure Chemical Co. Ltd) with average size 23 μm with 99.9% purity and copper powder (CUSP50) which with average particle size 4.8-5μm where used for second set of experiments.

2.2. Powder grinding

It is well known that efficient tool for grinding and mixing of many materials into fine powder is a ball milling process. There are two ways of grinding: the dry process and the wet process. The fine dispersion of the sub-stoichiometric copper-graphite composite powder successfully was fabricated by ball milling technique. The dry process of ball milling with using 5-6 mm size aluminum balls was carried out during 24 hours at room temperature. In the beginning of milling process, the powders and ball are mixed, and then they impact the inside surface of metallic bowl. Then the aluminum balls separated from mixed powder by sifting in thin metallic sieve.

2.3. Spark plasma sintering

The consolidation of all obtained powders was performed using spark plasma sintering (SPS) technique (Dr. Sinter 1030, Sumitomo Coal Mining Co. Ltd., and Japan) [15-17]. The schematic diagram of spark plasma sintering apparatus is shown in Fig. 1. In the first and second of the experiments were used different weight ratio of copper and graphite powders (amount will be described below). The mixture of copper and graphite powders were filled inside the mold of 20 mm diameter and the carbon paper was used to separate the powders from upper and lower punches. The total amount of powder was 8-10 grams. During the consolidation of powders at spark plasma sintering, heating rate was 200^{0}C/min and the pressure was 40MPa respectively. The samples were divided into groups in dependent of powder composition and sintering was done different temperature, varied 750-900^{0}C, respectively, during 5 minutes under flowing Ar-4%H$_2$ gas atmosphere. The obtained copper-graphite composites samples have approximately size 20mm in diameter and 4mm in thickness. Summarizes the different carbon-graphite samples synthesized by SPS and sintering condition (temperature, holding time and pressure) are shown in the Table 1.

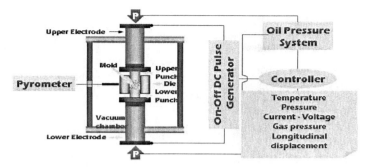

Fig. 1. Schematic diagram of spark plasma sintering apparatus.

Table 1: Summary of the sample composition (carbon and graphite powder (KB-3) with average size 30-40nm and 600 mesh graphite powder with average size 23 μm) and used experimental sintering condition (T-temperature; P-pressure; t_{hold} - holding time).

№	Sample name	Composition (wt%) KB-3 carbon & graphite	SPS T(°C)	P(MPa)	t_{hold}(min)	Sample name	Composition (wt%) 600 mesh graphite	SPS T(°C)	P(MPa)	t_{hold}(min)
1	Cu-C&Gr0.5	99.5%Cu+0.5%C&Gr	750	40	5	Cu-Gr0.5	99.5%Cu+0.5%Gr	750	40	5
2	Cu-C&Gr0.5	99.5%Cu+0.5%C&Gr	800	40	5	Cu-Gr0.5	99.5%Cu+0.5%Gr	800	40	5
3	Cu-C&Gr0.5	99.5%Cu+0.5%C&Gr	850	40	5	Cu-Gr0.5	99.5%Cu+0.5%Gr	850	40	5
4	Cu-C&Gr0.5	99.5%Cu+0.5%C&Gr	900	40	5	Cu-Gr0.5	99.5%Cu+0.5%Gr	900	40	5
5	Cu-C&Gr1	99%Cu+1%C&Gr	750	40	5	Cu-Gr1	99%Cu+1%Gr	750	40	5
6	Cu-C&Gr1	99%Cu+1%C&Gr	800	40	5	Cu-Gr1	99%Cu+1%Gr	800	40	5
7	Cu-C&Gr1	99%Cu+1%C&Gr	850	40	5	Cu-Gr1	99%Cu+1%Gr	850	40	5
8	Cu-C&Gr1	99%Cu+1%C&Gr	900	40	5	Cu-Gr1	99%Cu+1%Gr	900	40	5
9	Cu-C&Gr2	98%Cu+2%C&Gr	750	40	5	Cu-Gr2	98%Cu+2%Gr	750	40	5
10	Cu-C&Gr2	98%Cu+2%C&Gr	800	40	5	Cu-Gr2	98%Cu+2%Gr	800	40	5
11	Cu-C&Gr2	98%Cu+2%C&Gr	850	40	5	Cu-Gr2	98%Cu+2%Gr	850	40	5
12	Cu-C&Gr2	98%Cu+2%C&Gr	900	40	5	Cu-Gr2	98%Cu+2%Gr	900	40	5
13	Cu-C&Gr4	96%Cu+4%C&Gr	750	40	5	Cu-Gr4	96%Cu+4%Gr	750	40	5
14	Cu-C&Gr4	96%Cu+4%C&Gr	800	40	5	Cu-Gr4	96%Cu+4%Gr	800	40	5
15	Cu-C&Gr4	96%Cu+4%C&Gr	850	40	5	Cu-Gr4	96%Cu+4%Gr	850	40	5
16	Cu-C&Gr4	96%Cu+4%C&Gr	900	40	5	Cu-Gr4	96%Cu+4%Gr	900	40	5

2.4 Composite characterization

Copper-graphite composite were characterized in terms of mechanical, electrical, structural properties. The effect of heat treatment and powder composition on the mechanical properties of the composites was studied by measuring the electrical resistivity, bulk density and the micro-hardness of the specimens. The micro-hardness values of each polished specimens were measured 5 times by Vickers's micro-hardness tester (MMT-X3A), and the average value was obtained for each sample. This test method consists of indenting the test material with a diamond indenter, in the form of a right pyramid with a square base and an angle of 136 degrees between opposite faces subjected to different loads. The two diagonals of the indentation left in the surface of the material after removal of the load are measured using a microscope and their average calculated [18-19]. The bulk density of the sintered composites were measured by the water immersion technique (Archimedean method) by METTLER Toledo AG-204 devise using different balance and taking the temperature effect on the water density into account. The maximum achievable densities estimated based on the rule of mixture. The electrical resistivity of specimens was measured using high performance Hall System (HL 5500PC). The Hall measurement, carried out in the presence of a magnetic field, yields the sheet carrier density and the bulk carrier density if the conducting layer thickness of the sample is known. The Hall voltage for thick, heavily doped samples can be quite small (of the order of microvolts). In measurement were used four leads are connected to the four ohmic contacts on the sample. It is important to use the same batch of wire for all four leads and all four ohmic contacts should consist of the same material in order to minimize thermoelectric effects. The resistivity data was obtained by using standard calculation [20-21]. The morphology of the surface of the samples, polished down to 1µm diamond paste were determined by optical microscopy (Nikon L150) equipped with digital camera Nikon Digital Sight DS-U1 and scanning electron microscopy (JSM-6500F, Japan) and the chemical content of copper and carbon of the specimens was evaluated by EDS.

3.   RESULTS AND DISCUSSION
      Copper-graphite composites were prepared using graphite and copper as reinforcement by SPS at different temperature. Two types of graphite, with different composition, morphology and grain size were used: carbon and graphite powder (KB-3) with average size 30-40nm and 600 mesh graphite powder with average size 23µm. The differences in morphological structure and distribution of graphite particles at different composition are shown by the optical image in Fig. 2(a)-(e). Images of the obtained structures show distribution of graphite on copper composite, where darker areas represent higher graphite content and yellow areas represent higher content of copper.

3.1. Effect of the type of graphite on the structure of the copper-graphite composite.
      As the appearance of the as-received copper-graphite composite under the optical microscope images are shown in the Fig. 2(a)-(b) and (e)-(d)). These images show dissemination of graphite for each copper-graphite composition sintered at temperature 850^{0}C. The graphite disseminations represented as dark area and have not strong correlation for distribution graphite powder with different particles sizes. In particular, 600 mesh carbon-graphite powder with average size 23µm (see Fig. 2(c)) were observed in comparison with carbon- graphite powder (KB-3) which have average size 30-40nm (see Fig. 2 (g)) where each sample have the same content 98wt%:2wt%. In all cases, the graphite dispersion in copper structure was homogeneous with exception of small agglomerates for both 0.5wt% and 1wt% of graphite content (see Fig. 2 (a-b) and (e-f)). Probably, this is caused by uniform mixing powder during a ball milling process. The mainly difference was observed when graphite content was increased up to 4wt%. The samples, for instance, with 4wt% content of carbon graphite powder (KB-3) with average size 30-40nm were observed accumulation of the large size (~120µm) the graphite agglomerations (Fig. 2 (d)), while for graphite powder with average size 23µm the agglomerates were observed less frequently (Fig. 2 (h)) and their size also smaller (50µm). Probably the large agglomeration of the graphite with uniform distribution in the copper bulk improve mechanical properties, i.e. make a good solid lubricant and decrease the friction coefficient and wear rate of the material surface [22-24]. In fact, amount and size of agglomerations in the composition is directly related to graphite content and its powder size. A possible explanation for the graphite agglomerated structure might be the fact that the fine particles size of graphite lead to irregular dissolution of graphite in the Cu matrix and probably would have detrimental effect on the mechanic properties. Consequently their electrical properties also diminish strongly with type of graphite composite [16, 25].

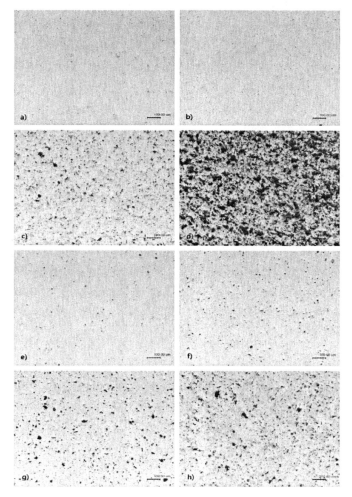

Fig. 2. Representative optical images of sintered samples : for copper and KB-3 carbon & graphite composite at T=850^{0}C: (a) Cu-C&Gr0.5; (b) Cu-C&Gr1; (c) Cu-C&Gr2; (d) Cu-C&Gr4; for copper and 600 mesh graphite composite at T=850^{0}C: (e) Cu-Gr0.5; (f) Cu-Gr1; (g) Cu-Gr2; (h) Cu-Gr4

Table 2. Micro-hardness, volume density of the copper graphite samples with different composition of graphite. Samples sintered by spark plasma sintering at 750^0C, 800^0C, 850^0C and 900^0C during 5 minutes under flowing Ar-4%H$_2$ gas atmosphere at pressure 40MPa respectively.

Sample name KB-3 carbon & graphite	Mechanical Properties			Sample name 600 mesh graphite	Mechanical Properties		
	Volume density, (g/cm^3)	Resistivity, (ohm-cm)x10^{-3}	Hardness, (HV)		Volume density, (g/cm^3)	Resistivity, (ohm-cm)x10^{-3}	Hardness, (HV)
Cu-C&Gr0.5	8.586	1.965	81.92	Cu-Gr0.5	8.550	2.017	77.78
Cu-C&Gr0.5	8.585	1.979	77.68	Cu-Gr0.5	8.562	1.998	73.5
Cu-C&Gr0.5	8.579	1.896	77.32	Cu-Gr0.5	8.551	1.936	68.5
Cu-C&Gr0.5	8.585	1.988	73.98	Cu-Gr0.5	8.545	1.952	67.96
Cu-C&Gr1	8.214	1.960	78.82	Cu-Gr1	8.447	1.993	77.42
Cu-C&Gr1	8.231	1.929	81.46	Cu-Gr1	8.445	1.942	72.48
Cu-C&Gr1	8.267	1.901	80.94	Cu-Gr1	8.430	2.017	75.58
Cu-C&Gr1	8.289	2.003	83.44	Cu-Gr1	8.417	2.0	70.84
Cu-C&Gr2	7.308	1.921	49.28	Cu-Gr2	8.029	1.906	77.9
Cu-C&Gr2	7.347	1.907	53.78	Cu-Gr2	8.036	2.0	76.64
Cu-C&Gr2	7.384	1.961	46.3	Cu-Gr2	8.048	1.909	75.62
Cu-C&Gr2	7.45	1.908	52.38	Cu-Gr2	8.046	1.959	75.4
Cu-C&Gr4	6.059	1.445	33.08	Cu-Gr4	7.730	1.983	65.4
Cu-C&Gr4	6.135	1.454	33.04	Cu-Gr4	7.749	1.894	74.5
Cu-C&Gr4	6.171	1.439	35.66	Cu-Gr4	7.780	1.917	75,76
Cu-C&Gr4	6.212	1.481	36.36	Cu-Gr4	7.781	1.962	75,92

The average densities of the as-received copper graphite composite with content of carbon-graphite powder (KB-3) with average size 30-40nm determined by the water immersion technique were table 2: 8.58 g/cm^3, 8.25 g/cm^3, and 6.144 g/cm^3, for the 0.5%, 1%, 2% and 4% respectively. The measured average densities for the 0,5%, 1%, 2% and 4% content of 600 mesh graphite powder with size 23μm are equal 8,55 g/cm^3, 8,43 g/cm^3, and 7,76 g/cm^3, respectively. The graphite content in samples is strongly affected to average densities. As expected, the density of graphite powder with size 23μm decreases slightly than graphite powder with average size 30-40nm (see Fig. 3 (a)) and keep this regularity with increasing of carbon graphite content for both types of graphite materials.

Hardness testing is the other most frequently used method for characterization mechanical properties of composite. An indenter of well-defined round shape geometry is pressed into surface of sample under predefined load. Therefore the micro-hardness indentations are micron size scale and have been extensively applied at the microstructure when Vickers diamond pyramid indenter is usually applied [19]. The hardness values for all copper-graphite specimens are given in table 2. The results show that maximum hardness value was obtained for the 1% of contamination of carbon-graphite

30nm powder. The results also show that increase of content for carbon-graphite powder (KB-3) with average size 30-40nm the hardness greatly decreased while hardness for the graphite powder with size 23µm is almost constant Fig. 3 (b).

One of the important characteristics of contact composite material is electrical resistivity. The electrical resistivity's of sintered copper-graphite composition at different content of graphite powder were calculated from resistance measurement of the Hall mobility and shown in Fig. 3 (c).

The copper-graphite samples with carbon and graphite powder (KB-3) which have average size 30-40nm (see Fig. 3 (c)) have resistance curve for different composition is similar for different contain of graphite. However, it was observed when used graphite powder with relatively big size 23µm the resistance curve start decrease at 2% and 4% containing of graphite powder. The decreasing of the electrical resistivity of composite samples with big and small powder size of both phases is shown in Fig. 3 (c and d). This dependence was remained even at higher sintering temperature. In other words increasing sintering temperature up to 900 ^{0}C for carbon-graphite powder (KB-3) with average size 23µm is almost linear distribution while carbon-graphite powder (KB-3) with average size 30-40nm the resistance curve greatly decreased Fig. 3 (d). This is caused that resistivity of both metallic powder and carbon pieces in matrix composite strongly dependent also from the shape of the carbon structure (big or small layer of carbon sheets, long or short fibers, spherical particles, etc.) and their structure orientation with respect to the current flow [4,8,16].

The morphology of the as-received samples is investigated by optical microscopic and FESEM images. Images of the obtained structures show distribution of graphite on copper composite, where darker areas represent higher graphite content and brighter areas represent higher content of copper. Especially for graphite KB-3 Carbon & Graphite powder with size 30-40nmthe dispersion in copper composite was homogeneous with uniformly distribution of graphite at low graphite content Fig.4-(a). After increasing 1wt% graphite content was observed homogeneous graphite distribution on composite surface which consist of the small black agglomerates Fig 4 (b). The magnification SEM image for 1wt% graphite content show that these black agglomerations are disseminated graphite islands Fig 4 (c) which proved by measurement of EDS spectrum Fig 4 (d).

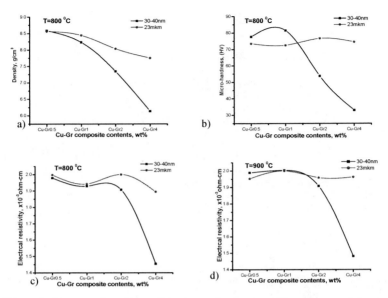

Fig. 3. Effects of of the type of graphite particle size on the structure of the copper-graphite composite, electrical and mechanical properties: (a) volume density, (b) micro-hardness, (c) resistivity of samples sintered at 800 ^{0}C, (d) resistivity of samples sintered at 900 ^{0}C

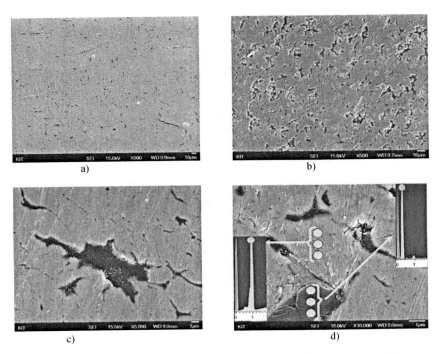

a)　　　　　　　　　　　　　　　　b)

c)　　　　　　　　　　　　　　　　d)

Fig. 4. FE-SEM image and EDS spectrum for copper graphite: a) copper-graphite composite with 0.5wt% gaphite ; b) copper-graphite composite with 1wt% gaphite; c) and d) are highy magnification image of copper-graphite composite with 1wt% gaphite and EDS spectrum of this sample, respectively.

4.　CONCLUSION

In this paper we report on the fabrication copper-graphite composite using spark plasma sintering technique and obtained samples investigated by micro-hardness, hall system, density tester, and optical microscope measurement methods. The effect of the ratio graphite content and powder size on the morphological, mechanical and electrical properties where investigated. In particular, as initial materials for the first type of the samples was used spherical copper powder with average particle size 4.8-5μm and carbon and graphite powder (KB-3) with average size 30-40nm. Initial materials for the second type of samples were used graphite powder with average size 23μm and same copper powder. The consolidation of all obtained powders was performed using spark plasma sintering technique. During the consolidation of powders at spark plasma sintering, heating rate was $200^0C/min$ and the pressure was 40MPa respectively. The samples were divided into groups in dependent of powder composition and sintering was done different temperature, varied 750-900^{0}C, respectively, during 5 minutes under flowing Ar-4%H$_2$ gas atmosphere. The conclusion is drawn as follows:

- The density of graphite powder with size 23 μm decreases slightly than powder with average size 30-40 nm and keep this regularity with increasing of carbon graphite content for both types of graphite materials. At the same time, micro-hardness of copper-graphite composites decrease gradually with increased powder size, in particular, of content for carbon-graphite powder (KB-3) with average size 30-40nm the hardness greatly decreased while hardness for the graphite powder with size 23μm is almost constant.

- It was observed decreasing the electrical resistivity of copper-graphite composite samples with different particle size. It is shown that increasing particle size and graphite content lead to reducing of the samples resistance. While for the carbon graphite powder (KB-3) with average size 30-40nm resistance almost similar.

- According the images optical microscope the surface morphology of the as-received copper-graphite composite is dependent from graphite content and sintering temperature. Besides it was found when graphite content was increased up to 4% for samples with carbon graphite powder (KB-3) with average size 30-40nm were observed accumulation of the large size (~120μm) the graphite agglomerations while for graphite powder with average size 23μm the agglomerates were observed less frequently and their size also smaller (50μm).

- Copper-graphite composite with 0,5wt% and 1wt% and particle size 30-40nm show higher density, micro-hardness and lower porosity than other composite copper-graphite materials with different contents of graphite.

REFERENCES

1. K. Rajkumar, K. Kundu, S. Aravindan, M.S. Kulkarni. "Accelerated wear testing for evaluating the life characteristics of copper–graphite tribological composite", *Materials and Design 32* (2011) 3029–3035
2. T. Futamia, M. Ohirab, H. Mutoa, M. Sakaia, "Contact/scratch-induced surface deformation and damage of copper–graphite particulate composites", *CARBON 47*, (2009), pp. 2742-2751
3. J.M. Garcia-Marquez, N. Anton, A. Jimenez, M. Madrid, M.A. Martinez and J.A. Bas, "Viability study and mechanical characterisation of copper–graphite electrical contacts produced by adhesive joining", *J. Mater. Process. Technol. 143/144* (2003), pp. 290–293
4. D.H. He, R. Manory and H. Sinkis. " Effect of alloying elements on the interfacial bonding strength and electric conductivity of carbon nano-fiber reinforced Cu matrix composites". *Materials Science and Engineering A 449/451*, (2007), pp. 778–781.
5. S.G. Sapate, A. Uttarwar, R.C. Rathod, R.K. Paretkar, "Analyzing dry sliding wear behavior of copper matrix composites reinforced with pre-coated SiCp particles", *Materials and Design 30*, (2009), pp.376–386.
6. A. Guillet, E.Y. Nzoma, P. Pareige, "A new processing technique for copper-graphite multifilamentary nano-composite wire: Microstructures and electrical properties", *Journal of Materials Processing Technology 182*, (2007), pp. 50–57.
7. K.F. Mayerhofer, E. Neubauer, Ch. Eisenmenger-Sittner, H. Hutter, "Characterization of Cr intermediate layers in Cu-C-system with SIMS method", *Applied Surface Science* 179 (2001) 275-280.
8. T. Futamia, M. Ohirab, H. Mutoa, M. Sakaia, "Indentation contact behavior of copper–graphite particulate composites: Correlation between the contact parameters and the electrical resistivity", *CARBON 46*, (2008), pp. 671-678.
9. Q. Hong, M. Li, J. Wei, et al., "New Carbon-Copper Composite Material Applied in Rail-Type Launching System", *IEEE TRANSACTIONS ON MAGNETICS*, VOL. 43, NO. 1, JANUARY 2007.

10. D.H. He, R. Manory, H. Sinkis, "A sliding wear tester for overhead wires and current collectors in light rail systems", *Wear 239*, (2010), pp. 10-20.
11. G. Bucca, A. Collina, "A procedure for the wear prediction of collector strip and contact wire in pantograph–catenary system", *Wear 266*, (2009), pp. 46–59.
12. S.G. Jia, P. Lui, F.Z. Ren, et al., "Sliding wear behavior of copper alloy contact wire against copper-based strip for high-speed electrical railways", *Wear 262*, (2007), pp. 772-777
13. S.F. Mustafa, E.I. El-Badry, A.M. Sanad, "Effect of graphite with and without copper coating on consolidation behavior and sintering of copper-graphite composite", *Powder Metallurgy* 1997; 40(3), pp. 201-6.
14. S.F. Mustafa, E.I. El-Badry, A.M. Sanad, B. Kieback, "Friction and wear copper-graphite composites made with Cu-coated and uncoated graphite powders", *Wear 253*, (2002), pp. 699-710
15. S. J. Kim etc."The Fabrication of Porous Hydroxyapatite Including of Nano-Sized TiN_x by Spark Plasma Sintering", *7th Pacific Rim Conference on Ceramic and Glass Technology*, Nov14 (2007)
16. J.M. Ullbrand, J.M. Cardoba, J.T. Ariztondo, "Thermomechanical properties of copper-carbon nanofibre composites prepared by spark plasma sintering and hot pressing", *Composites Science and Technology 70*, (2010), pp. 2263-2268.
17. S. J. Kim, Y. H. Oh, S. B. Park "Fabrication of Sputtered $TiO_{2-x} N_x$ Photocatalyst Using Spark Plasma Sintered Targets", *Materials Science Form*, Mar 30, (2007), pp. 539-543.
18. A. Shakeel, Shahdad, F.J. McCabe, et al., "Hardness measured with traditional Vickers and Martens hardness methods", *Dental Materials 23*, (2007), pp. 1079-1085
19. L. Sidjanin, D. Rajnovic, J. Ranogajec, E. Molnar, "Measurement of Vickers hardness on ceramic floor tiles", *Journal of the European Ceramic Society 27*, (2007), pp. 1767–1773
20. "Test Methods for Measuring Resistivity and Hall Coefficient and Determining Hall Mobility in Single-Crystal Semiconductors", *ASTM Designation F76*, Annual Book of ASTM Standards, Vol. 10.04 (2011).
21. R. Chwang, B. J. Smith and C. R. Crowell, "Contact Size Effects on the van der Pauw Method for Resistivity and Hall Coefficient Measurement," *Solid-State Electronics 17*, (1974), pp. 1217-1227.
22. M. Kestursatya, J.K. Kim and P.K. Rohatgi, "Wear performance of copper-graphite composite and a leaded copper alloy", *Mat. Sci. Eng. A339* (2003), pp. 150–158.
23. J. Kovacik, S. Emmer, J. Bielek, et al., "Effect of composition on friction coefficient of Cu–graphite composites", *Wear 265*, (2008), pp. 417–421
24. K. Rajkumar, S. Aravindan," Microwave sintering of copper–graphite composites", *Journal of Materials Processing Technology 209*. (2009), pp. 5601–5605.
25. A. Kumar, M. Kaur,R. Kumar, et al., "Development of pitch-based carbon–copper composites", *Journal Material Science 45*, (2010), pp. 1393–1400.

DENSIFICATION AND MICROSTRUCTURE CHANGES OF CERAMIC POWDER BLENDS DURING MICROWAVE SINTERING

Audrey GUYON, Jean-Marc CHAIX, Claude Paul CARRY, Didier BOUVARD
Laboratoire SIMaP, Grenoble INP / CNRS / UJF
BP46 38402 Saint Martin d'Hères, France

ABSTRACT

A comparative study between microwave and conventional sintering was carried out on alumina-zirconia composites with different volume fractions of zirconia. The microwave sintering experiments were achieved in a single-mode resonant cavity (2.45 GHz) in a direct configuration heating, without susceptor, at constant incident power and under predominant electric field. The results obtained in terms of densification and microstructure changes are compared to the ones obtained in case of conventional sintering, using the same thermal cycles. This work shows that 10 vol.% of zirconia in alumina powder are enough to allow direct microwave heating of composite blend and that the microwaves lead to an enhancement of densification. The microstructures show a grain size gradient due to thermal gradients, which suggests that the microwave sintering temperatures have been slightly underestimated. Nevertheless, these thermal gradients are not sufficient to totally explain the decrease of the sintering temperatures, which should thus be due to a microwave effect.

INTRODUCTION

Microwave heating differs fundamentally from conventional heating methods since the use of microwaves enables energy transfer directly to materials, allowing bulk, selective and ultrafast heating[1-3]. Microwave sintering offers several advantages compared to conventional sintering methods: ultrafast heating rate contributing to significantly reduced cycle time and increased densification rates at relatively low sintering temperatures leading to finer microstructures and improved properties, all these specificities promoting low energy consumption[1-5].

Microwave sintering has proved its applicability on dielectric materials as ceramics[6,7] that can couple with the oscillating electric field. Most authors showed that ceramic powders can be sintered by microwaves at lower temperatures and for shorter times than in conventional sintering, which allows for finer and more uniform microstructures to be obtained. Some of them claimed the existence of a genuine "microwave effect", *i.e.* the acceleration of diffusion mechanisms by the oscillating electric field, which should explain the enhancement of the sintering process[8].

In the past many studies dealing with conventional sintering of alumina and zirconia have been carried out, which involved the characterization of the behavior of these materials with both macroscopic (densification) and microscopic (microstructures) view during classical heating[4,5,9]. The benefits of microwave heating led numerous researchers to investigate the behavior of these materials under microwaves[1,4,10-12]. These studies, among others, highlighted the importance of material dielectric properties, especially the dielectric loss factor, with respect to the interaction between the electromagnetic field and the material[1,5,12]. Indeed, as alumina presents poor dielectric losses over a range of temperature from room temperature to 1000°C, its heating by microwaves is difficult at low temperatures[5]. Several solutions are possible to overcome this drawback. It is possible to use an external preheating system (usually called susceptor and made of a material with high dielectric losses such as silicon carbide)[2-5] or to add to the initial powder a material with higher dielectric losses that interacts with microwaves at low temperatures[5]. Zirconia, responding to these features[4,10], could be such internal susceptor. Furthermore alumina-zirconia composites are interesting for various applications.

Many studies have been achieved on microwave sintering of alumina-zirconia composites [2,3,5]. However, the diversity of experimental conditions (single-mode or multimode cavity, direct heating or use of a susceptor, thermal insulation…), which are not always clearly specified in the literature, and the difficulty of reliable temperature measurement complicate the comparison of microwave experiments with each other or with conventional sintering tests. The interest of the work presented here is linked to the use of an original microwave sintering furnace for the realization of experiments with the best possible control of process parameters. This set-up includes a single mode resonant cavity, allowing thus a control of the electromagnetic field distribution in the cavity and so a control of the position of the sample with regard to the electric and magnetic fields. This study concerns direct microwave heating, i.e., without susceptor, of alumina-zirconia composites with higher alumina fraction. In the following, the microwave furnace is described and the results of sintering experiments are presented. For each material, the effects of conventional and microwave heating on densification and microstructures are discussed.

EXPERIMENTAL PROCEDURE

Materials
    Different commercial powders, provided by the Baikowski Society (France), have been used for this study. The studied compositions include: pure non-doped alpha alumina (BMA15, S_{BET} = 16 ± 2 m²/g, d_{BET} = 94 nm, ρ_{th} ≈ 3,95 g/cm³), pure yttria stabilized zirconia with 3mol.% of yttrium oxide (Y-ZrO$_2$, S_{BET} = 32 ± 2 m²/g, d_{BET} = 31 nm, ρ_{th} ≈ 6,09 g/cm³) and two blends of alumina and zirconia containing 40 and 10 vol.% of Y-ZrO$_2$, respectively. Both mixtures have been obtained by wet attrition of both previously described ultrafine pure powders. The green compacts are achieved by uniaxial compaction (50 MPa) followed by cold isostatic pressing (100 to 300 MPa depending on the powder) in order to obtain cylindrical samples of about 8 mm by 8 mm and with an initial density between 50 and 60%. The applied pressure must be tailored according to the powder since each one has a specific compressibility. At last, specimens are subjected to a binder removal under air, with a heating rate of 120°C/h until 600°C following by a holding time of 3h, to remove all organic agents by pyrolysis (weight loss of about 1 to 3% depending on the powder).

Sintering Experiments
    Conventional and microwave sintering experiments were carried out using the same thermal cycle in order to compare both heating techniques: 25°C/min until the sintering temperature without holding time. The selected heating rate is a compromise between classical sintering (the highest heating rate that the dilatometer can follow) and microwave sintering (low heating rate compared to those achievable). Classical sintering experiments were performed under air flow in a vertical dilatometer (SETSYS Evolution TMA, SETARAM France) while microwave sintering experiments were carried out under air flow in a single-mode resonant cavity (2,45 GHz[4]). The experiment setup used for microwave heating is schematized in Figure 1. A rectangular wave-guide of section 86.36 x 43.18 mm² allows for the transport of the microwave radiation to a rectangular TE$_{10p}$ cavity. This resonant cavity is ended by a coupling iris (a vertical slot in a copper sheet) on the magnetron side and by a reflector, also called sliding short-cut piston, on the other side. Samples are set upon a quartz plate inside a vertical 40 mm-diameter quartz tube that goes across the cavity through two holes and that can be filled with various atmospheres. To reduce microwaves escape, a 93 mm high chimney is fixed to the cavity at each hole. Microwave heating being characterized by a bulk heating in a cold environment (the cavity), the specimen is surrounded by an insulating material, transparent to microwaves, in order to limit heat losses through the surface and thus reduce the thermal gradients (from the bulk to the surface of the sample)[1,5,11]. The temperature of the upper surface of the specimen is continuously

measured through a hole in the insulating material by a thermal imaging camera located above the centre of the cavity and protected by a ZnSe window.

During microwave heating, the forward electromagnetic wave is reflected against the conductive wall of the short-cut piston, leading to a standing wave made of a single stacking of forward and reflected waves. In the empty cavity, the points where the magnetic and electric fields are respectively maximum and minimum (near to zero) are exactly distant from each other by a quarter of the guided wave-length (43 mm). Besides, when the iris-piston distance (cavity length) is equal to 4 or 5 times the half guided wave-length, the iris reflects almost entirely the wave in the cavity (TE_{104} or TE_{105} resonant mode). The resonance phenomenon consists in the stacking of multiple waves reflected against the piston and the iris, which results in an increase of the electromagnetic energy in the cavity. In our cavity, the TE_{104} and TE_{105} resonant modes are used for materials interacting respectively with the magnetic field or the electric field. Ideally, the iris-specimen distance should be chosen to be about equal to half of the cavity length that leads to a resonant mode. That allows the best coupling between the material and the electric field at resonance (maximum dissipated power). However we are aware that the simple electric and magnetic patterns expected in an ideal – close and empty – cavity are significantly perturbed by the chimneys and the sintering compact. In particular it is most likely that the compact never undergoes pure magnetic field or pure electric field. Anyway, we assume that in TE_{104} (respectively TE_{105}) configuration the magnetic field (respectively electric) is predominant.

In this study, the microwave sintering experiments were carried out under predominant electric field in a direct configuration heating ; this means there was no susceptor and the sample was only surrounded by a fibrous insulating material composed of alumina and silica. Finally, the heating rate was controlled by adjusting manually the short-cut piston position at constant incident power (500 W) and keeping the iris fixed[4]. We checked that pure alumina powder could not be heated up to a temperature high enough to allow sintering.

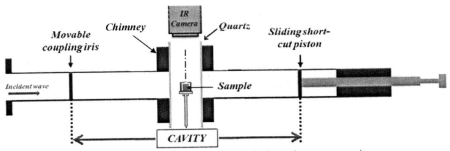

Figure 1. Schematic of the microwave single mode resonant cavity.

Characterization

The weight density of the samples was measured by the Archimedes method using ethyl alcohol. The microstructure of the sintered materials was studied on polished and thermal etched specimens using a scanning electron microscope (SEM) equipped with field emission gun (Zeiss ULTRA-55 FEG SEM, with Schottky field emission source, CMTC-Grenoble INP). The average grain size of each phase deduced from SEM observation was determined using the average intercept method.

RESULTS AND DISCUSSION

Conventional sintering

The densification curves obtained during classical heating at 25°C/min heating rate for the four different powders (pure alumina, pure yttria doped zirconia and the two powder blends) are plotted in Figure 2. These curves evidence that the densification behaviors of pure alumina and zirconia powders are close to each other until about 1300°C. Above this temperature alumina densifies slightly faster than zirconia. In addition, the densification curves of alumina-zirconia mixtures are shifted by about a hundred degrees towards high temperatures compared to the ones of pure powders. The mixture with 10 vol.% of Y-ZrO$_2$ shows a slightly slower densification than the mixture with a higher content of zirconia.

This apparent retardation of sintering in case of the addition of zirconia particles in alumina, characterized by higher sintering temperatures, has also been reported by J. Wang and R. Raj[13]. They showed that only 5 vol.% of zirconia in alumina leads to higher sintering temperatures than pure alumina. They linked this delay in terms of sintering to an increase of the sintering apparent activation energy without giving any information about the micro-mechanisms responsible for this increase. Nevertheless, we suggest that small zirconia grains (30 nm), acting as an inert second phase for alumina grains (90 nm), can lead to a blocking "mechanical" effect (10 vol % of zirconia is equivalent to 3 zirconia grains for 1 alumina grain) resulting in the shift of the densification curves toward high temperatures. In the case of the 10 vol.% Y-ZrO$_2$, the low amount of ZrO$_2$ inclusions, acting as an inert second phase, postpone significantly the sintering of alumina to higher temperatures than those of pure materials. The case of the 40 vol.% Y-ZrO$_2$ is different because each material forms a percolating network. Although both materials sinter approximately at the same rate (Figure 2, pure alumina and pure zirconia densification curves), they hinder each other, which once again slows down the densification. The outcome that 40 vol.% Y-ZrO$_2$ is slightly less detrimental to the densification than 10 vol.% seems to be fortuitous.

The microstructures of powder blends after classical sintering are presented in Figure 3. They look like those obtained by Samuels et al.[5]. Both materials exhibit equiaxed grains with an homogeneous distribution of zirconia in alumina at the considered scale (the composition fluctuations observed on SEM micrographs at on images at mag. 20 000 on the 40 and 10 vol.% Y-ZrO$_2$ are typically +/-5% and +/-2%, respectively). Also we checked that the grain size was homogeneous within the specimens. The average grain size of alumina in 40 vol.% Y-ZrO$_2$ mixture is slightly lower. J. Wang and R. Raj[13] have shown that growth of the alumina grains is significantly impeded by addition of only 5 vol.% of zirconia in the final stage of sintering (relative density higher than 0.9). Thus, it is expected that a higher amount of zirconia has a more important effect on the alumina grain growth rate.

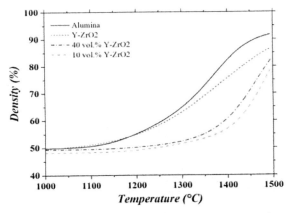

Figure 2. Densification curves during conventional sintering at 25°C/min heating rate for the different studied powders: pure Al$_2$O$_3$, pure Y-ZrO$_2$ and the Al$_2$O$_3$-ZrO$_2$ blends with 10 and 40 vol.% of Y-ZrO$_2$.

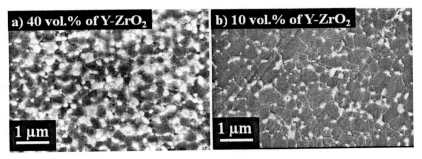

Figure 3. SEM microstructures of alumina-zirconia composites after conventional sintering at 1500°C and 25°C/min: a) 40 vol.% of Y-ZrO$_2$ and b) 10 vol.% of Y-ZrO$_2$.

Microwave sintering

We know that alumina presents poor dielectric losses at low temperatures which does not allow direct microwave heating as reported by R. Raj et al.[14]. The experiments performed in our microwave furnace showed that only 10 vol.% of dispersed zirconia allow the heating of non-lossy alumina. We assume that zirconia particles are heated by coupling with the microwaves and they heat alumina particles through conduction or radiation. Alumina particles can thus reach a temperature high enough to permit their direct coupling with the microwaves up to the temperature required for sintering. The microwave sintering of this two-phase material is an example of the great influence of a dispersed minor phase on microwave processing of a powder compact as reported by R. Raj et al.[14].

The densification curves obtained during conventional sintering are compared with the densities of composite specimens sintered by microwave heating at various temperatures on Figure 4. Whatever the amount of zirconia was, a significant difference between conventional and microwave heating can be observed. Thus, it appears that microwave sintering allows decreasing the sintering temperatures for a targeted relative density (Figure 4). Indeed, the obtained data show that, for a same final relative density (95% for instance), the sintering temperature required in microwave heating is lower than the one required in conventional heating whatever the composition of the mixture was. This beneficial effect of microwaves on sintering resulting in an acceleration of the densification and therefore in a decrease of the sintering temperatures has also been reported by R. Raj et al.[14] et by several other authors[4,5,10]. R. Raj et al.[14] proposed that this beneficial effect can be due to a higher dielectric loss at grain boundaries than the one in the grain matrix. Thus the local temperatures at the grain boundaries should be significantly higher than that of the average temperature of the sample, leading to a faster rate of diffusional transport. The study of the possible microwave effect on the densification mechanisms of alumina-zirconia composites is yet so far in progress. At last, the effect of microwaves seems to be more pronounced when the zirconia content is lower since it leads to a higher decrease of the sintering temperatures. This last point must be confirmed by the study of alumina with lower content of zirconia.

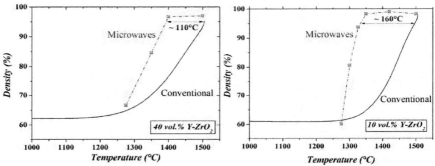

Figure 4. Densification of alumina-zirconia powder mixtures during direct microwave and conventional sintering at 25°C/min : 40 vol.% Y-ZrO$_2$ (left) and 10 vol.% of Y-ZrO$_2$ (right).

SEM microstructures obtained after microwave and conventional heating in the bulk of samples for both powder blends are presented in Figure 5. The microstructures obtained after microwave sintering are similar to those obtained after conventional sintering for either the 40 vol.% powder blend or the 10 vol.% one (Figure 5). Indeed, each sample shows equiaxed alumina and zirconia grains with a uniform distribution of zirconia in alumina. Nonetheless, the grain size of microwave-sintered samples seems to be higher whatever the blend was, for the same sintering conditions. To clarify this point the average grain size of each phase for the 40 vol.% composite has been estimated for both microwave and conventional heating experiments. In order to detect possible microstructure heterogeneities, we measured the grain size at various heights within each sample. The obtained data are plotted in Figure 6. The average grain size of both alumina and zirconia phases in the upper part of the microwave-sintered sample is identical to the one in the upper part of conventionally-sintered sample. But, the grain size along the axis of the conventionally-sintered sample is almost constant for either alumina or zirconia phase whereas it varies along the height of the microwave-sintered sample. Nevertheless, the grain size in case of microwave sintering is almost constant between the bottom and the middle parts and reduces towards the top surface of the specimen. This suggests that microwave sintering induces a grain size gradient. But this gradient may be also due to a thermal gradient. The hole in the top part of the insulating material for temperature measurement can lead to higher thermal losses and induce a thermal gradient in the densifying compact. This kind of microstructure gradient, due to thermal gradients, has been underlined by Mizuno et al.[11]. These observations suggest that the temperature measured on the top surface is slightly lower than the one in the bulk of samples. The real temperature in the bulk of samples is thus underestimated and so the effect of microwaves on the decrease of sintering temperatures is overestimated. Thus, there are two reasons for the observed higher grain size in the microwave-sintered sample: underestimated temperature and earlier densification. In order to clarify both last points, other experiments with different sintering temperatures are in progress.

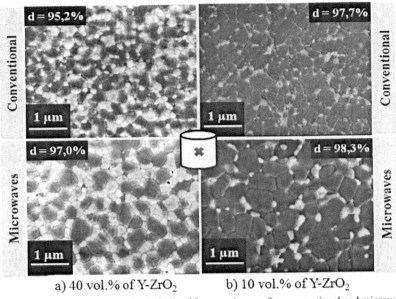

a) 40 vol.% of Y-ZrO$_2$　　　b) 10 vol.% of Y-ZrO$_2$

Figure 5. SEM microstructures in the bulk of four specimens after conventional and microwave sintering at 1500°C and 25°C/min for alumina-zirconia powder blends containing 40 vol.% of Y-ZrO$_2$ (a) and 10 vol.% of Y-ZrO$_2$ (b).

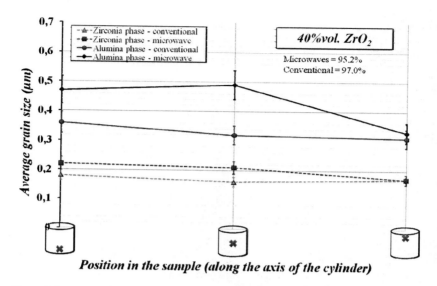

Figure 6. Average grain size for both alumina and zirconia phases after direct microwave and conventional sintering at 1500°C and 25°C/min of alumina – 40 vol.% zirconia powder blend.

CONCLUSION

We investigated direct microwave sintering of mixtures of non-lossy alumina and lossy zirconia powders. We found that 10 vol.% of zirconia particles in the alumina powder is enough to allow direct microwaves heating with almost full densification at 1350°C, zirconia particles acting thus as an internal susceptor. The comparative study of conventional and microwave sintering evidenced an enhanced densification under microwave heating for every powder mixtures, which means that theses mixtures can be sintered at lower temperature. Nevertheless, the microstructures of microwave sintered samples revealed a grain size gradient along the axis of the samples, likely due to thermal gradients. This observation suggests that the temperature in microwave heating has been underestimated. It means that the enhancement of densification is not as pronounced as it has been deduced from densification measurement. However this enhancement is so strong that it can not be fully explained by the underestimate of the temperature.

REFERENCES

[1]S. Das, A. Mukhopadhyay, S. Datta, and D. Basu, Prospects of microwave processing: an overview, *Bull. Mater. Sci.*, **31**, 946-956 (2008).

[2]R.R. Menezes, and R.H.G.A. Kiminami, Microwave sintering of alumina-zirconia nanocomposites, *J. Mater. Process. Tech.*, **203**, 513-517 (2008).

[3]N. A. Travitzky, A. Goldstein, O. Avsian, and A. Singurindi, Microwave sintering and mechanical properties of Y-TZP/20 wt.% Al$_2$O$_3$ composites, *Mat. Sci. Eng. A.*, **286**, 225-229 (2000).

[4]S. Charmond, C. P. Carry, and D. Bouvard. Densification and microstructure evolution of y-tetragonal zirconia polycrystal powder during direct and hybrid microwave sintering in a single-mode cavity, *J. Eur. Ceram. Soc.*, **30**, 1211-1221 (2010).

[5]J. Samuels, and J. R. Brandon, Effect of composition on the enhanced microwave sintering of alumina-based ceramic composites, *J. Mater. Sci.*, **27**, 3259-3265 (1992).

[6]T. T. Meek, C. E. Holcombe, and N. Dykes, Microwave sintering of some oxide materials using sintering aids, *J. Mater. Sci. Lett.*, **6**, 1060-1062 (1987).

[7]D. E. Clark, and W. H. Sutton, Microwave processing of materials, *Annu. Rev. Mater. Sci.*, **26**, 299-331 (1996).

[8]J. Binner, K. Annapoorani, A. Paul, I. Santacruz, and B. Vaidhyanathan, Dense nanostructured zirconia by two stage conventional/hybrid microwave sintering, *J. Eur. Ceram. Soc.*, **28**, 973-977 (2008).

[9]G. Bernard-Granger, and C. Guizard, Apparent activation energy for the densification of a commercially available granulated zirconia powder, *J. Am. Ceram. Soc.*, **90**, 1246-1250 (2007).

[10]A. Goldstein, N. Travitzky, A. Singurindy, and M. Kravchik, Direct microwave sintering of yttria-stabilized zirconia at 2,45 GHz, *J. Eur. Ceram. Soc.*, **19**, 2067-2072 (1999).

[11]M. Mizuno, S. Obata, S. Takayama, S. Ito, N. Kato, T. Hirai, and M. Sato, Sintering of alumina by 2,45 GHz microwave heating, *J. Eur. Ceram. Soc.*, **24**, 387-391 (2004).

[12]M. Oghbaei, and O. Mirzaee, Microwave versus conventional sintering: A review of fundamentals, advantages and applications, *J. Alloy. Compd.*, **494**, 175-189 (2010).

[13]J. Wang, and R. Raj, Activation energy for the sintering of two-phase alumina/zirconia ceramics, *J. Am. Ceram. Soc.*, **74**, 1959-1963 (1991).

[14]R. Raj, M. Cologna, and J. S. C. Francis, Influence of externally imposed and internally generated electrical fields on grain growth, diffusional creep, sintering and related phenomena in ceramics, *J. Am. Ceram. Soc.*, **94**, 1941-1964 (2011).

DENSIFICATION OF UO$_2$ VIA TWO STEP SINTERING

J. Vidal
AREVA, AREVA NP GmbH
Erlangen, Germany
Ecole Nationale Supérieure de Céramique Industrielle, GEMH
Limoges, France

M. Zemek
AREVA, AREVA NP GmbH
Erlangen, Germany

P. Blanchart
Ecole Nationale Supérieure de Céramique Industrielle, GEMH
Limoges, France

ABSTRACT

In this work, Two Step Sintering (TSS) was used aiming to obtain high densification of nuclear fuel compacts with a more homogeneous structure, which could be beneficial to the thermo-mechanical properties and in-reactor operational behavior. In TSS, it is possible to favor either grain growth or densification by heating the sample to an optimal temperature above which the grain growth accelerates dramatically. It is followed by a cooling to a lower temperature and soak for a prolonged time in order to favor further densification with limited grain growth. To verify the applicability of TSS to UO$_2$ fuel compacts, two types of powders were used with different particle sizes: - UO$_{2.25}$ powder (particle size of about 260 μm), which underwent successively reduction and oxidation stages; - UO$_{2.07}$ as dry converted powder (particle size of about 120 μm). TSS was performed using either an oxidizing in CO$_2$ or a reducing atmosphere in H$_2$. Temperature optimization was performed with dilatometric studies of green compacts. For each studied powder, an optimized TSS process was attained and studied in details. Microstructural characteristics of sintered compacts were characterized. It has been shown the ability of TSS to achieve a density higher than 95%TD in H$_2$ atmosphere with an average grains size of about 1 μm.

INTRODUCTION

The control of microstructural characteristics and particularly the limitation of excessive grain growth during sintering[1] of nuclear fuel ceramics is very important for the in-reactor operational behavior. Irradiation induces defects such as vacancies, interstitials and fission atoms, with extended defects such as bubbles, pores and dislocations. They move at grain boundaries, which act as sinks for defects of all types. In highly irradiated UO$_2$, fission gas induces the formation of intra granular bubbles, which favor swelling[2].

In nanomaterials, the high proportion of boundaries to bulk is favorable to viscoplasticity and to resistance of irradiation damage[3]. The microstructural characteristics and mainly grain and pore size distribution determine the fuel performances[4, 5], and mostly the retention of fission gas at interfaces. To control the microstructural characteristics, both the adjustment of thermal cycles and a deep understanding of involved sintering mechanisms must be achieved. The occurrence of different sintering mechanisms during the final stage of sintering ensures that densification and grain growth are competitive processes[6]. During this sintering stage, an accelerated grain growth process occurs when the pinning effect of pores or second phases is no longer efficient due to pore collapse[7] or to the change of morphology and distribution of second phase[8].

173

In recent years, Chen and Wang[9] were the pioneers of an effective method for the sintering of ceramics, called the two-step sintering (TSS), favoring densification, but with minimal grain coarsening. Many authors have already applied successfully this sintering process with different oxide ceramics, such as Al$_2$O$_3$ [10, 11, 12], BaTiO$_3$ [13], La$_2$O$_3$ [14], SiC [15], TiO$_2$ [16], Y$_2$O$_3$ [17, 18] and ZrO$_2$ [19, 20, 21]. In a first step, the sample is heated to a high temperature in order to reach a significant but not too high intermediate relative density (0.75 to 0.85, depending on the oxide). According to Chen and Wang[9], volume diffusion mechanism is predominant so grain coarsening occurs. In a second stage, it is cooled down and maintained at a lower temperature to process a long time isothermal sintering until full density is attained. Since supercritical pores disappear during the first step, remaining pores become unstable and the sample can subsequently sinter at a lower temperature. A key parameter is the control of the relative density at the end of the first step, to maintain a sufficiently high density of triple point junctions[22]. It is determinant to suppress the grain boundary movement during the second sintering step since only grain boundary diffusion remains active to reach a high density of the material without significant grain coarsening.

With oxides ceramics, the kinetics and the thermodynamics of sintering can be drastically changed as a function of sintering conditions, such as oxygen partial pressure (pO$_2$) of the sintering atmosphere[23] and the characteristics of starting powders (specific surface area...). With UO$_2$, many of these parameters were studied by Dörr and Assmann[24]. They evidenced that UO$_2$ pellets sintered at about 1200°C during a short time and under small pO$_2$ (10^{-3} to 10^{-5}Pa) attain a high relative density of 95%, similar to that of pellets conventionally sintered in hydrogen (pO$_2$ in the range of 10^{-9} to 10^{-11}Pa) at 1750°C. Atmosphere type changes the mechanism of sintering, since in a reducing atmosphere the low concentration of uranium vacancies induces a small driving force for sintering and hence a 1700°C sintering temperature. In more oxidizing atmosphere, the O/U ratio is increased. That favors a large density of uranium vacancies and a high value of uranium diffusion coefficient [25, 26]. Beside the control of microstructural characteristics, this sintering technique is favorable to reduce energy consumption and equipment costs, and to increase the production rate thanks to the relatively short thermal cycle.

According to the reported works about TSS process applied to oxide ceramics and using the existing knowledge relative to UO$_2$ sintering, new experiments were performed to determine the appropriate thermal cycle of TSS, in both oxidative and reducing atmospheres, thus at low and high temperatures respectively. The present study will attempt to validate the TSS process for the reduction of residual porosity in UO$_2$ material.

EXPERIMENTAL

Two UO$_2$ powders were used, which were obtained through the dry conversion process (DC):
- UO$_{2+x}$ with x adjusted at 2.25 by the addition of U$_3$O$_8$. U$_3$O$_8$ was prepared by heating the DC-UO$_2$ powder in air at 400°C for four hours and then by passing through a 90 μm sieve to remove large agglomerates. The sieved U$_3$O$_8$ powder has an average particle size of 70 μm.
- UO$_2$ as calcined powder which underwent successively five oxidation and reduction stages (called UO$_2$ RedOx).

The main characteristics of the UO$_2$ powders used in this study are given in Table I.

Table 1: Characteristics of the UO$_2$ powders.

	O/U	Specific surface (BET) [m²/g]	Mean particle size* [μm]
UO$_2$ as calcined	2.07	2.35	120
UO$_2$ RedOx	2.25	3.73	260

* Laser granulometry by dry dispersion [27].

To facilitate the shaping by compaction of green pellets, 0.2wt% of Cirec (N, N' Bi-stearamide) was added as lubricant before the pressing of green pellets. For this study, green pellets were prepared following successive steps:
- sieving the mixed powder through 150 μm and 90 μm sieves
- homogenization of the mixture in a Turbula-mixer
- precompaction as slugs (density of 3.9 g/cm³)
- granulating of the pre-compacts (screen size of 1.18 mm)
- addition of Cirec to the granulates and homogenization in a Turbula-mixer
- biaxial pressing of the granulated powders into cylindrical green pellets (height 9.5 mm, diameter 8.1 mm), with a green density of 53%±1% of theoretical density.

All sintering studies were performed using two different dilatometry equipments, to precisely record densification curves and to determine the optimal sintering conditions (temperature, atmosphere, sintering stage…). A typical dilatometry curve of UO$_{2.25}$ sintered in hydrogen (H$_2$) is presented in Fig. 1. It presents the density variation and the shrinkage rate as a function of temperature. At about 480°C the reduction of UO$_{2+x}$ to UO$_2$ occurs and it is followed by a peak of densification rate at 1600°C.

In a preliminary study, sintering experiments of different fuel types were performed in order to determine the most accurate T1 temperature of the TSS process under oxidative and reducing atmospheres. For all thermal cycles, a constant heating rate up to the final temperature was used. Heating of fuel samples was firstly performed in a reducing atmosphere (H$_2$) up to 1750°C without holding time and immediately cooled down to room temperature. It will be called High Temperature Sintering (HTS). In a slightly oxidative atmosphere (CO$_2$), samples were also heated at the lower temperature of 1200°C during three hours, and this process will be called Low Temperature Sintering (LTS). At 1200°C, CO$_2$ atmosphere was maintained during two hours and replaced by H$_2$ for additional 1 hour, to adjust the final material stoichiometry. For both sintering processes, pellets were heated up and cooled down by a rate of 10°C/min. The thermal cycles are shown schematically in Fig. 2.

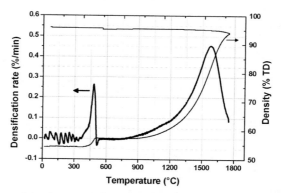

Fig. 1: Densification and density curves with temperature of hyperstoichiometric UO$_2$ sintered under H$_2$.

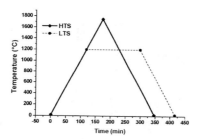

Fig. 2: Thermal cycles of preliminary sintering processes at high (T1) and low temperatures (T2) in reducing and oxidative atmospheres respectively.

The TSS process was performed with a first heating at T1 temperature, at which a maximum shrinkage rate of the material was obtained during preliminary tests (Fig. 1). After T1 achievement, a decreasing step of 100°C temperature[9] was performed down to T2 temperature. At T2, samples were sintered during a period of 12 hours[22] in oxidative or reducing atmosphere. Oxidative sintering at T2 was followed by a dwell time of one hour in H$_2$, to control the stoichiometry. The initial heating rate and final cooling rate were 10°C/min, while intermediate cooling rates (from T1 to T2) were 20°C/min and 30°C/min in reducing and oxidative atmospheres respectively.

Sintered densities of UO$_2$ pellets were measured by the water immersion method[28] and reported as a percentage of the theoretical density of UO$_2$ (10.96 g/cm³). The pellets were subsequently cut and polished and to reveal the grain boundaries, chemical etching was applied on the polished pellets. The sintered compacts were examined to characterize porosity, size and morphology of grains. Both optical microscopy and scanning electron microscopy (SEM) were used. The average grain size was determined by the lineal intercept method and pore size distribution was obtained by the Saltykov[29] correction of results.

RESULTS

The influence of T1 temperature was studied during preliminary experiments. Table II shows the relative sintered density (D). All compacts exhibit a final density higher than 95% theoretical density (TD) indicating the good ability to sinter under the two studied atmospheres. In general, grains sizes are larger in oxidative atmosphere, due to the increased diffusion rate of uranium.

Table II: Sintering of $UO_{2.25}$ and UO_2 RedOx.
GS: Grain size of sintered samples; $T\varepsilon max$: temperature of maximum densification rate; $D\varepsilon max$: relative density at $T\varepsilon max$.

	Sintering	D (%TD)	GS (μm)	$T\varepsilon max$ (°C)	$D\varepsilon max$ (%TD)
$UO_{2.25}$	H_2-1750°C	95.8	3	1590	84
UO_2 RedOx	H_2-1750°C	97.8	5	1410	85
$UO_{2.25}$	CO_2 (+1h H_2)-1200°C-2h	97.2	6	940	78
UO_2 RedOx	CO_2 (+1h H_2)-1200°C-2h	96.1	10	900	68

In the present work, T1 temperatures were determined considering the variation of the shrinkage rate $d(\Delta L/L0)/dt$ during the preliminary sintering. T1 temperatures ($T\varepsilon max$) are reported in Table II against the density of the compact when the maximal shrinkage rate occurs ($D\varepsilon max$).

The shrinkage peak of the dilatometric curves is attained when density ranges from 78 to 84%TD for hyperstoichiometric UO_2, and from 68 to 85%TD for UO_2 RedOx. This behavior depends also on the sintering atmosphere and the densification rate below the peak temperature. Particularly, the densification rate of UO_2 RedOx powder is faster than that of $UO_{2.25}$ due to the higher specific surface area (Fig. 3). The T1 temperature will be used during the TSS process under the two sintering atmospheres.

Two-step sintering results are reported in Table III. It reports the conditions under which TSS regimes were performed and the final characteristics of the pellets after different sintering processes. The effect of four sintering parameters on density and grain size were tested: T1, T2, holding time at T1 and heating rate from the intermediate T' temperature (end of debinding) below T1. Density at T1 temperature is also tabulated.

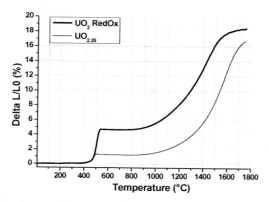

Fig. 3: Dilatometric curves of preliminary sintering process at high temperature (1750°C) in reducing atmosphere.

Table III: Sintered densities and grain sizes of UO$_2$ compacts under various parameters of TSS.

	Atmosph.	T' (°C)	T1 (°C)	D at T1 (%TD)	Dwell time at T1 (min)	T2 (°C)	D (%TD)	GS (μm)
				Sintering conditions				
UO$_{2,25}$	CO$_2$		940	78		840	92.4	0.8
			1100	91		900	94.7	1.4
	H$_2$		1590	84		1490	96.6	3.2
			1650	91		1350	93.5	0.9
		800	1650	89		1450	96.3	1.3
UO$_2$ RedOx	CO$_2$	800	900	78		800	90.6	0.8
			1100	93	10	900	94.3	3.0
	H$_2$		1410	85		1310	95.1	0.8
		900	1410	81		1310	95.5	0.9

Typical microstructures of hyperstoichiometric UO$_{2.25}$ and RedOx UO$_2$ are presented in Fig. 4, Fig. 5 and Fig. 6 for ceramics sintered in H$_2$. To characterize the pore size distribution (Fig. 4 and Fig. 5), polished surfaces were observed with the optical microscope. Samples were also chemically etched and observed by SEM (Fig. 6), to determine the grain size of compacts.

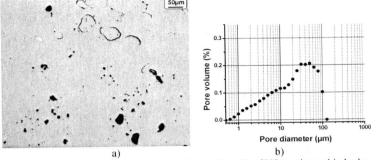

a) b)

Fig. 4: Porosity microsection (a) and pore size distribution (b) of $UO_{2.25}$ sintered in hydrogen (green sample firstly heated up to T'=800°C, then to T1=1650°C, cooled down to T2=1450°C and held at T2 for 12 hours).

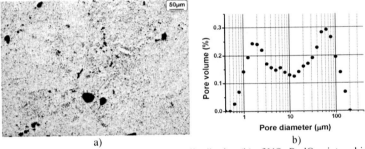

a) b)

Fig. 5: Porosity microsection (a) and pore size distribution (b) of UO_2 RedOx sintered in hydrogen (green sample firstly heated up to T'=900°C, then to T1=1410°C, cooled down to T2=1310°C and held at T2 for 12 hours).

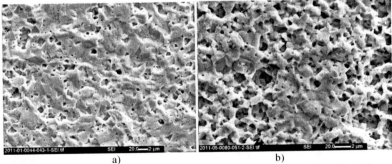

a) b)

Fig. 6: SEM images of UO_2-U_3O_8 with a 1.3 μm grain size (a) and UO_2 RedOx with a 0.9 μm grain size (b).

DISCUSSION

Preliminary sintering experiments result in relatively low T1 temperatures, in ranges of 1410 to 1590°C (H$_2$ atmosphere) and of 900 to 940°C (CO$_2$ atmosphere), in comparison to the maximum sintering temperatures of 1700°C and 1200°C usually used. At T1 temperature, relative densities (Table II) attain adequate values reported by Chen and Wang[9], which claim that a sufficiently high starting density should be reached during the first step. However, the value of density varies according to authors. Chen[9] indicated that the critical density is greater than 75% for Y$_2$O$_3$, while Li[30] and Bodisova[12] reported values of 82% and 92% for respectively submicron and nanometric powder compacts of alumina. They all agree that the sample should be sintered to a sufficiently high temperature, to attain a relative density at which all pores are unstable and can be removed during shrinkage.

During TSS in CO$_2$, the influence of T1 temperature is evidenced in Table III. Whereas the large variation of T1 temperature in the range of 900 to 1100°C, the relative density at the completion of TSS is always lower than 95%TD, and this observation is valid for the two used powders. For hyperstoichiometry UO$_2$, the maximum density (94.7%TD) is attained for T1=1100°C, but it is still too low to activate a sufficient densification at T2.

In H$_2$ atmosphere, T1 is in the range of 1410 to 1590°C. The relative density is between 80% and 85% for both powders. The relative density at the completion of TSS is always above 95%TD. It means that at T1, the surface energy is still high to ensure that free energy of the system leads to dense compacts during the second stage[31].

The influence of T2 is also reported in Table III. T2 values were selected according to Chen et al.[9], who considered for Y$_2$O$_3$ that T2 must be 100°C below T1. In this study, the authors carried out the first tests with this T2-T1 difference. Samples sintered in H$_2$ attain a high relative density, but when the final density exceeds 96%TD (UO$_{2.25}$), the grain size increases significantly (3.2 μm). In CO$_2$, the same 100°C difference leads to very low relative density, below 93%TD. When T2-T1 increases to 200°C, it is favorable to a high densification with a small grain size of UO$_{2.25}$. It is related to both the high T1 and T2 temperatures under H$_2$ atmosphere, since a 300°C difference is unfavorable to a high densification. Under CO$_2$, the T2-T1 difference of 200°C does not lead to similar results since the driving force for sintering due to uranium vacancies is still high at low temperature [25], leading an increased grain size.

The roles of the heating rate up to T1 and of the dwell time at T1 were also studied. Values were determined taking account results of Bodisova[12] with alumina samples. Most of our experiments were performed using a predetermined rate of 10°C/min up to T1, without any dwell time at this temperature. However, it is a too rapid rate for the achievement of chemical equilibrium and the complete dewaxing. Slower heating rates can prevent defects such as microcracks that are induced during the Cirec extraction below 900°C. Consequently a series of experiments with the two powders were conducted with a first 5°C/min heating rate up to T' temperature of 800°C or 900°C, followed by a 20°C/min or 30°C/min heating rate up to T1.

UO$_{2.25}$ and UO$_2$ RedOx compacts sintered in H$_2$ attain high relative density (Table III), close to the target. For UO$_{2.25}$ the value of 96.3%TD is attained with an average grain size of 1.3 μm, which evidences the determinant role of fast heating rate combined with an optimal sintering temperature. Moreover, this variation of heating rate before T1 leads to decrease the density at T1, as it is observed with UO$_2$ RedOx (with hyperstoichiometric compact as well), but it does not have any significant consequence on the final density and grain size.

In oxidative CO$_2$ atmosphere, a UO$_2$ RedOx sample was sintered from T' to T1 with a heating rate of 30°C/min. This high heating rate was determined from preliminary results of UO$_2$ RedOx sintering at high temperature. Using heating rates of 10°C/min and 20°C/min up to the same T1 and T2 did not change significantly the density and grain size. For materials sintered at 1100°C in CO$_2$, these results suggested to use a faster heating rate of 30°C/min and an additional isothermal dwell of 10 min. It results in a relative density of 94.7%TD, but with an increased grain size of 3 µm.

Micrographs of hyperstoichiometric $UO_{2.25}$ and UO_2 RedOx sintered in H_2 are presented in Fig. 4 and 5, respectively. Photos of the two samples reveal a very broad distribution of pore size with irregular forms. Size attains 100 µm in $UO_{2.25}$ and 110 µm in UO_2 RedOx. The presence of larger pores is supposed to be favored by Cirec decomposition and evaporation during heating. A particle size distribution analysis of this additive by laser granulometry revealed particle size in a similar range of larger pores (20 to 100 µm). For $UO_{2.25}$, whereas the relative density attains 96.3%TD, a small amount of large pores is observed (Fig. 4b), probably from inhomogeneities in the UO_2-U_3O_8 powder compact. Large pores are formed in some grain boundaries and a coalescence phenomenon due to the preferential diffusion along grain boundaries is clearly noticeable. These large pores might be obstacles to grain boundary migration, so grain growth can be impeded. For UO_2 RedOx (Fig. 5b) a higher fraction of medium pore size, below 5 µm, induces a slightly lower density after sintering (95.5%TD).

$UO_{2.25}$ in Fig. 6a shows an average grain size of 1.3 µm in a homogeneous and dense grain microstructure (96.3%TD) but no abnormal grain size is observed. Since the pellet has deviated from the stoichiometry, a preferential densification of agglomerates is observed on the microstructure. UO_2 RedOx in Fig. 6b shows dense zones with a pore size of about 2 µm. They are surrounded by a more porous material with a higher density of interconnected pores and small grains (<1 µm).

Similar results of densification and grain growth interaction as a function of the characteristics of the powder compacts were often extensively discussed in the literature. Particularly, Rahaman[32] and Kang[33] reviewed the effect of pores on grain growth during the final stage of conventional sintering. They evidenced that grain growth is changed by remaining pores that pin the grain boundaries, which cannot migrate. Porosity was also considered by Fang[34] as a second phase, which inhibits boundary motion. A similar behavior occurs in the microstructure of UO_2 RedOx, where pores prevent the excessive grain growth during the final sintering stage. Grain size does not exceed about 900 nm in a relatively dense material (95.5%TD). But the mechanisms of pore coarsening must be studied in more details to prevent the formation of very large pores observed in Fig. 4a and Fig. 5a. Beside the role of inhomogeneities in the powder compact, several mechanisms were proposed as the coalescence of fine pores under grain boundary dragging and grain boundary diffusion[35].

From all results, Fig. 7 presents how the TSS process is favorable to high density with small grain size, since adequate sintering parameters are used. The results must be compared to that of usual sintering processes, which results to a drastic increase of grain size with relative density during the last sintering and densification stage. But for UO_2, since T1, T2, dwell times and heating rates are very important parameters, the sintering atmosphere is also determinant in densification and grain growth processes. Fig. 7 illustrates that H_2 atmosphere is preferable for the densification rate and leads to density higher than 96% with a small grain size below 2 µm.

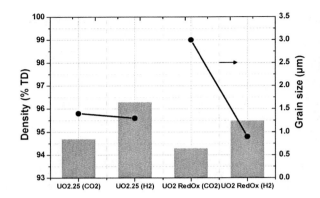

Fig. 7: Conclusive results of TSS under reductive and oxidative atmospheres for $UO_{2.25}$ and UO_2 RedOx powders.

CONCLUSIONS

The TSS process is a new way to obtain UO_2 sintered ceramics with an optimized density up to 96%TD. But the adequate microstructural characteristics can be obtained only by the precise adjustment of thermal cycle temperatures and times, as by the atmosphere type during sintering stages. Beside, the powder characteristics, the stoichiometry of the initial UO_{2+x} phase and the green compact characteristics are also very important.

In TSS, the T1 first stage temperature is a key parameter for the control of the intermediate relative density. For $UO_{2.25}$ and UO_2 RedOx, a relative density in the range of 85 to 90%TD is appropriate for both sintering atmospheres. Beside density, the determination of necessary microstructural characteristics at the end of the T1 stage should be examined as well.

The second-stage temperature T2 is also a very important parameter to ensure that the final relative density attains the highest possible value without an excessive grain growth. During the T2 stage, the predominant densification mechanism shall be grain boundary diffusion, to slow down the grain growth phenomenon. With UO_2, the authors showed that T1-T2 temperature difference should exceed in the literature recommended value of 100°C, to attain 200°C.

The T2 stage is also accompanied by an excessive pore coarsening induced by inhomogeneities in the starting powder compacts and by fine pore coalescence from grain boundary diffusion. It leads to the occurrence of very large pores which contribute to limit the upper value of the sintered density.
The study continues with further optimization of the temperature program and/or applying grain growth inhibitor additions to UO_2.

ACKNOWLEDGMENTS
The authors thank AREVA NP GmbH Fuel Laboratory in Erlangen, Germany for the long time support of this research.

REFERENCES

[1]Y. Liu, B.R. Patterson, Grain growth inhibition by porosity, *Acta metal. Mater*, **41** (1993)

[2]I. Zacharie, S. Lansiart, P. Combette, M. Trotabas, M. Coster, M. Groos, Thermal treatment of uranium oxide irradiated in pressurized water reactor: Swelling and release of fission gases, *Journal of Nuclear Materials*, **255** (1998)

[3]G. Ackland, Controlling Radiation Damage, *Science*, **327** (2010)

[4]C.Y. Joung, S.C. Lee, Fabrication method for UO$_2$ pellets with large grains or a single grain by sintering in air, *Journal of Nuclear Materials*, **375** (2008)

[5]K.W. Song, D.S. Sohn, Effects of sintering processes on the duplex grain structure of UO$_2$, *Journal of Nuclear Materials*, **200** (1993)

[6]R. Marder, R. Chaim, Grain growth stagnation in fully dense nanocrystalline Y$_2$O$_3$ by spark plasma sintering, *Materials Science and Engineering*, **527** (2010)

[7]R.J. Brook, Controlled Grain Growth, in treatise on *Materials Science and Technology*, **9** (1976)

[8]K. Chang, W. Feng, L.Q. Chen, Effect of second-phase particle morphology on grain growth kinetics, *Acta Materialia*, **57** (2009)

[9]I.W Chen, Sintering dense nanocrystalline ceramics without final stage grain growth, *Nature*, **404** (2000)

[10]K. Maca, Two step sintering of oxide ceramics with various crystal structures, *Journal of the European Ceramic Society*, **30** (2010)

[11]Z. Razavi Hesabi, M. Haghighatzadesh, Suppression of grain growth in sub-micrometer alumina via two step sintering method, *Journal of the European Ceramic Society*, **29** (2009)

[12]K. Bodisova, P. Sajgalik, Two-stage-sintering of alumina with submicrometer grain size, *J. Am. Ceram. Soc.*, **90** (2007)

[13]X.H. Wang, P.L. Chen, Two-step-sintering of ceramics with constant grain size, II: BaTiO3 and Ni-Cu-Zn Ferrite, *J. Am. Ceram. Soc.*, **89** (2006)

[14]Y. Huang, D. Jiang, Fabrication of transparent lanthanum-doped yttria ceramics by combination of two step sintering and vacuum sintering, *J. Am. Ceram. Soc.*, **92** (2009)

[15]Y.I. Lee, Y.W. Kim, Effect of processing on densification of nanostructured SiC ceramics fabricated by two-step, *Journal of materials science*, **39** (2004)

[16]M. Mazaheri, Z.R. Hesabi, Two-step sintering of titania nanoceramics assisted by anatase-to-rutile phase transformation, *Scripta Materialia*, **59** (2008)

[17]M. Mazaheri, M. Valefi, Two step sintering of nanocrystalline 8Y$_2$O$_3$ stabilized ZrO$_2$ synthesized by glycine nitrate process, *Ceramics International*, **35** (2009)

[18]X.H. Wang, P.L. Chen, Two-step sintering of ceramics with constant grain size, I.: Y2O3, *J. Am. Ceram. Soc.*, **89** (2006)

[19]K. Maca, V. Pouchly, Two-step sintering of oxide ceramics with various crystal structures, *Journal of the European Ceramic society*, (2009)

[20]J. Tartaj, P. Tartaj, Two stage sintering of nanosize pure zirconia, *J. Am. Ceram. Soc.*, **92** (2009)

[21]C.J. Wang, C.Y. Huang, Two-step sintering of fine alumina-zirconia ceramics, *Ceramics International*, (2008)

[22]M. Mazaheri, A. Simchi, F. Golestani-Fard, Densification and grain growth of nanocrystalline 3Y-TZP during two-step sintering, *Journal of the European Ceramic Society*, **28** (2008)

[23]K.W. Lay, Role of the O/U ratio on the sintering of UO$_2$, *Journal of Nuclear Materials*, **30** (1969)

[24]Dörr, Assmann, Sintering of UO$_2$ at low temperatures, *Elsevier scient. Publ. Comp.*, Amsterdam, Oxford, New York (1980)

[25]T.R.G. Kutty, P.V. Hegde, K.B. Khan, U. Basak, S.N. Pillai, A.K. Sengupta, G.C. Jain, S. Majumdar, H.S. Kamath, D.S.C. Purushotham, Densification behaviour of UO$_2$ in six different atmospheres, *Journal of Nuclear Materials*, **305** (2002)

[26]T.R.G. Kutty, P.V. Hedge, Characterisation and densification studies on ThO$_2$-UO$_2$ pellets derived from ThO$_2$ and U$_3$O$_8$ powders, *Journal of Nuclear Materials*, **335** (2004)

[27]J. Vidal, M. Zemek, P. Blanchart, Particle size distribution of uranium dioxide powders by laser dry diffraction method, *Powder technology*, in review.

[28]International Standard ISO 9278:1992 (E), Uranium dioxide pellets – determination of density and amount of open and closed porosity – Boiling water method and penetration immersion method

[29]J.L. Chermant, M. Coster, Use of Saltykov Corrective Method with a Semi-Automatic and Automatic Image Analzers, *Praktische Metallurgie*, **14** (1977)

[30]J. Li, Y. Ye, Densification and grain growth of Al2O3 nanoceramics during pressureless sintering, *J. Am. Ceram. Soc.*, **89** (2006)

[31]P. Duran, F. Capel, A strategic Two-Stage Low-Temperature Thermal Processing leading to fully dense and fine-grained doped- ZnO varistors, *Advanced Materials*, **14/2** (2002)

[32]M.N. Ragaman, *Ceramic processing and sintering*, 2nd edition, (2003)

[33]S.-J. L. Kang, Sintering: densification, grain growth and microstructure, MA, Butterworth-Heinemann, (2005)

[34]Z.Z. Fang, Densification and grain growth during sintering of nanosized particles, *International Materials Review*, **53** (2008)

[35]H. Eckart Exnerw, C.Müller, Particle Rearrangement and Pore Space Coarsening During Solid-State Sintering, *J. Am. Ceram. Soc.*, **92** (2009)

EFFECT OF TWO-STEP SINTERING ON OPTICAL TRANSMITTANCE AND MECHANICAL STRENGTH OF POLYCRYSTALLINE ALUMINA CERAMICS

Hyung Soo KIM [a,b], Young Do KIM[b], and Sang Woo KIM[c]

[a]R&D Center, KINORI Co., Ltd., Sungnam Gyeonggi 462-806 Korea
[b]Division of Materials Science and Engineering, Hanyang University, Seoul 133-791 Korea
[c]Clean Energy Research Center, Korea Institute of Science and Technology, Seoul 136-650, Korea

ABSTRACT

The effect of two-step sintering (TSS) process on optical and mechanical strength of polycrystalline alumina ceramic was investigated. High purity alumina compacts with a ppm level of impurities were prepared by a ceramic injection molding technique for the two-step pressureless sintering under vacuum atmosphere. The first sintering step was carried out by a fast heating rate to reach higher sintering temperatures within a short time, followed by cooling and holding at a lower temperature for the subsequent second step. The temperature of 1780°C was effective for the first step sintering (FSS) showing the highest densification rate. The second isothermal sintering was then carried out at 1300–1600°C (TSS) for various hours in order to avoid the grain boundary migration and to enhance segregation of secondary phases in the grain boundary. The two-step sintering temperature and time increase, the grain size tends to gradually increase, which leads to promote homogeneous grain microstructure without any significant grain growth and to enhance transparency of polycrystalline alumina without degradation of mechanical strength.

1. INTRODUCTION

Polycrystalline translucent alumina has excellent transmittance of visible and infrared light, and also has excellent thermal, mechanical and electrical stability. In its early stage of development it was used in sodium lamps, metal-halide lamp, high-quality electronic parts, and high-frequency insulation, while more recently it has been widely used in dental supply parts due to its characteristics of aesthetics and strength.[1-4] Sapphire, a single crystalline alumina, is a high-density, high-purity material that is suitable for dental applications due to its excellent corrosion resistance, its harmlessness to human body and its high breaking strength.[5] However, it has the disadvantage of being difficult to embody with powder metallurgy and powder injection molding (PIM), and its processing cost is quite high.[6] Polycrystalline alumina material has a lower light transmittance than sapphire, but has about the same corrosion resistance and strength. Since the application of PM & PIM processes to polycrystalline alumina is feasible, much research is in progress in this field.[7,8]

The optical and mechanical properties of polycrystalline alumina ceramics are critically dependent on the secondary phases, impurities, and abnormal grains in the sintered bodies. To achieve high strength and transparency in the polycrystalline alumina, the impurity levels and micro-structural homogeneities should be controlled at atomic level. Segregated impurities or second phases that act as light scattering centers severely deteriorate optical and mechanical strength of the pressureless sintered bodies. The segregation of the impurities in intra-grains or grain boundaries can be controlled by elaborate sintering techniques.

In this research, the green body fabricated through the powder injection molding method was sintered through an elaborate two-step sintering method to observe the microstructure of the sintered body and measure its strength and light transmittance. For sintering, the temperature was increased to 1780°C, maintained at that temperature for a short period, and then lowered. In the second step, the temperature was maintained at 1300-1600°C for each different time to enable precise control of the grain size and grain size distribution.

2. EXPERIMENTAL METHOD AND PROCESS EQUIPMENT

2.1 Powder Characteristics

The feedstocks used in powder injection molding are composed of ceramic powder which determines the physical and chemical properties of products and binders to provide fluidity to the powder. High-purity alumina powder AKP-3000 (α-Al_2O_3, Sumitomo Chemical, Japan) generally used in CIM with a diameter less than 1 micron was used, and for high temperature sintering 500 ppm of magnesium oxide (MgO) was added as a sintering aid to control the grain size depending on sintering conditions.

2.2 Preparation of Mixtures and Injection Molding

For the bonding agent system used in the experiment, paraffin wax and polyacetal, self-developed thermoplastic multi-component polymers, were used as main materials. The optimum compactness of the mixtures for injection molding was determined through a mutual comparison of viscous behavior by capillary viscosity, mixture density and microstructure, and the optimum powder compactness of high purity alumina powder/ magnesium oxide powder mixture was 50 vol %. The mixture was prepared at 150°C using a double planetary mixer (Toshin, 3L, Japan) for one hour. The mixture was crushed after cooling, pelletized to enable easy injection molding, and then was injection-molded into a specimen mold (Disc-type 1.5 t*12Φ mm) which is used to measure 30 mm gauge length strength and transmission in an injection molding machine (Woo Jin, NA55) with a clamping force of 55 tons. In order to secure shape stability during injection molding, the structure of the side gate and the pin point gate applied in the middle of the disc was determined using the Moldflow Injection Molding Simulation Program for preparation and injection molding of the standard disc specimen mold in a way that can ensure the injection molding stability.

2.3 Polymer-binder Agent Removal Process

A two-step process composed of solvent extraction and follow-up thermal decomposition was used for the removal of binding agents. The solvent extraction was conducted by dipping the specimen into a solvent, and n-heptane was used as the solvent. The binding agents were extracted into the solvent at 40°C for 5 hours to remove more than 95 % by weight. The injection molded body after solvent extraction went through thermal decomposition under atmospheric condition. For thermal decomposition behavior, the body was heated at a rate of 1°C/min, maintained at 200~500°C for 2 hours, and then cooled at the outside.

2.4 Sintering Process

Sintering was conducted to remove the binders through thermal decomposition by using the tungsten sintering furnace, and the heating rate was 4°C/min. The grain size was controlled by sintering at high temperatures in a vacuum. The first step was to raise the temperature to the maximum of 1780°C to enable easy control of the grain size, maintain the temperature for a short period of time, and then lower it. The second step was to maintain the temperature between 1300 and 1600°C for each time interval to control the grain size growth.

2.5 Analysis of Properties after Sintering
2.5.1. Analysis of Light Transmittance

To measure the effects of sintering temperature and sintering time on the light transmittance, the above-mentioned sintered body specimen was processed to a size of 0.8 mm and then the light transmittance was measured over the visible light region of 380 nm ~780 nm using UV-IR Spectrometer (Scinco, S-4100, Korea).[9]

2.5.2. Test of Biaxial Flexural Strength

To measure the strength of the disc specimen used for the measurement of light transmittance, the biaxial flexural strength was measured using the ASTM F394-78 method, and the test method is shown in Fig. 1

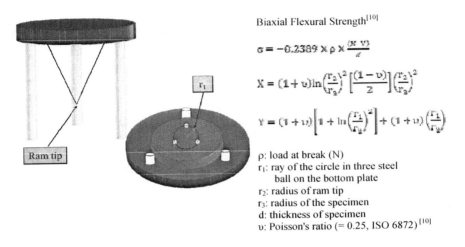

Biaxial Flexural Strength[10]

$$\sigma = -0.2389 \times \rho \times \frac{(X-Y)}{d}$$

$$X = (1+\upsilon)\ln\left(\frac{r_2}{r_3}\right)^2\left[\frac{(1-\upsilon)}{2}\right]\left(\frac{r_2}{r_3}\right)^2$$

$$Y = (1+\upsilon)\left[1+\ln\left(\frac{r_1}{r_3}\right)^2\right]+(1+\upsilon)\left(\frac{r_1}{r_3}\right)$$

ρ: load at break (N)
r_1: ray of the circle in three steel
  ball on the bottom plate
r_2: radius of ram tip
r_3: radius of the specimen
d: thickness of specimen
υ: Poisson's ratio (= 0.25, ISO 6872)[10]

Fig. 1. Schematic diagram of the biaxial flexural strength fixture.

2.5.3. Analysis of Microstructure

For the microstructure of the sintered body, the polished cutting surface of the sintered body was thermally etched at 1500°C and then observed with SEM[6] , but since the temperature was higher than the isothermal heat treatment temperature of 1300~1600°C for the two step of this experiment, it was difficult to observe grain size change. Thus, the surface of the sintered body injection-molded with a high-polishing injection mold and sintered was observed with a Scanning Electron Microscope (Hitachi; X-4200) and the grain size and grain size distribution were analyzed through the ASTM-E112 method using the image analysis program (innerview m-series). At this time, the surface of the specimen was coated with Pt using a sputter, and observation was made at an acceleration voltage of 15 kV.

2.5.4. Numerical simulation for injection molding process.

A standard mold with disc mold sides, centered side gate and centered pin point gate was assumed, and the disc-injection-molded specimen was simulated for the injection molding process using the Moldflow program (Moldflow, MPI 6.0)

3. RESULT & DISCUSSION

3.1 Comparison of Injection Molding based on Mold Gate Method

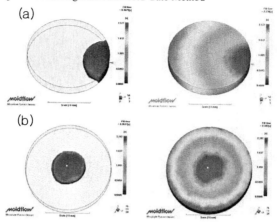

Fig. 2. Simulated results showing stress distributions developed during ceramic injection molding process of powder-polymer blend mixtures with different mold gates; (a) normal side gate, (b) pin-point centered gate.

The simulation results for injection molding of the standard disc specimen with the side gate and pin point gate confirm that in the mold with the side gate, the resin is filled directionally during the injection molding process, as shown in Fig. 2 (a). In this case, an internal stress is created due to left/right unbalanced filling characteristics of the injection molded body, and so shape deformation and cracks due to the difference in filling density are highly likely to take place during the post-molding processes of solvent extraction, binder removal by heating, and sintering. In the mold structure with centered pin point method as shown in Fig. 2(b), however, symmetrical filling is secured on the basis of the center, and so it is possible to minimize the deformation due to stress difference during the binder removal process.[11]

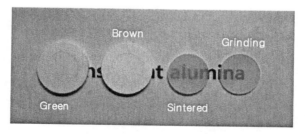

Fig. 3. Powder injection molded samples; as-molded (green), debinded (brown), and sintered bodies.

In addition, to improve the uniform filling balance and resin flow and minimize injection molding variables that may influence the light transmittance and strength characteristics, a 0.02 mm air-vent was added to the outside of the mold core during the fabrication of the mold, and it was confirmed that the sintered body without deformation could be made as shown in Fig. 3.

3.2 Grain size Analysis of Sintered Body (Scanning Electron Microscope)

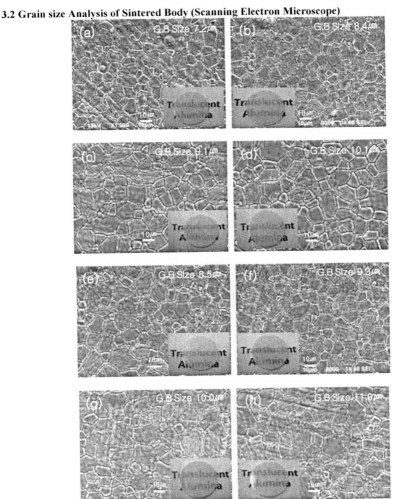

Fig. 4. SEM surface microstructure of specimens first-step sintered (FSS) at 1780°C for 5min in vacuum atmosphere; under various heating parameters of two-step sintered (TSS1_1300°C) at; (a) 1, (b) 5, (c) 10, and (d) 20 hr , (TSS2_1600°C) at; (e) 1, (f) 5, (g) 10, and (h) 20 hr

After maintaining specimens at the maximum temperature of 1780℃ for 5 minutes, specimens were heat-treated for the TSS second step at 1300℃ (TSS1) and 1600℃ (TSS2). The microstructure and grain size/grain size distribution of specimens at TSS conditions could be confirmed through SEM analysis, as shown in Fig. 4 and 5. Fig. 4(a) and Fig. 4(h) show the smallest (7.2 μm) and largest average grain size (11.0 μm), respectively. The growth rate of the grain sizes was still sluggish at T_2, indicating that the activation energy for grain-boundary diffusion is far lower than that for grain-boundary migration, thus, the second isothermal sintering process is promoted by grain-boundary controlled kinetics without fast grain growth.[12,13] Although the subsequent sintering step is ineffective to promote further densification in our experiment, it leads to promote homogeneous grain microstructures without any significant grain growth and to enhance a good distribution of second phases in the grain boundary. Therefore, it confirms that the sintering temperature and time in the second step have the greatest influence on the final grain microstructure. At both temperatures of TSS1 (1300℃) and TSS2 (1600℃), the grain size was confirmed to increase proportionally with increasing sintering time. Thus, the light transmittance of specimens could also be predicted to increase with increasing sintering time.

At the 1300℃/1hr conditions of Fig. 5(a) where the grain size was the smallest, the frequency of the average grain size below 5 μm was analyzed to be high, and the distribution narrow. At the 1600℃/20 hr conditions of Fig. 5(h), where the sintering temperature was high and the sintering time was the longest, there were grains above 20 μm in size, which shows that the grain growth took place as the sintering period increased. The results of the analysis of overall grain size distribution show that the average grain size tends to increase with increased sintering time at the same sintering temperature.

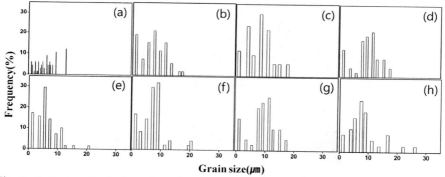

Fig. 5. Grain-size distribution of specimens first-step sintered at 1780℃ for 5 min in vacuum atmosphere; followed by 2-step sintering at 1300℃ for (a) 1, (b) 5, (c) 10, (d) 20 hr, and at 1600℃ for (e) 1, (f) 5, (g) 10, and (h) 20 hr.

Table 1 summarized the results of light transmittance at 780 nm wave length, biaxial flexural strength, sintered body density and average grain size for each TSS holding time at TSS temperatures of 1300 and 1600℃. For the TSS holding time of 1 hour there were no significant differences in light transmission between sintering temperatures, but for the TSS holding time of 20 hours the light transmittance increased with increased sintering temperature. In particular, at 1600℃/20 hr the light transmittance was higher than in the other three temperature conditions (1300℃~1500℃). At TSS1 (1300℃) conditions, the grain size increased proportionally with TSS holding time even though the rate of increase was small, but the proportional relationship could not be found for the remaining 3 measurements of light transmittance, strength and density.[6] This appears to be due to the grain size distribution. At TSS2 (1600℃) conditions both the light transmittance and grain size increased with an increase in TSS holding time, but beyond 5 hours the strength did not show much difference. This indicates that at the highest sintering temperature (1600℃) and the longest TSS holding time (20 hours), the grain size was more uniform than under other conditions. At these conditions, the highest light transmittance of 19.0 % (much higher than the rest) around the visible light region of 780 nm without degradation of mechanical strength could be confirmed.

Table 1. Summary of detailed physical properties of specimens two-step sintered at temperatures of 1300 and 1600℃

$T_1(℃)$ - $t_1{}^a$(min)	$T_2(℃)$- $t_2{}^b$(hr)	Optical transmittance(%)	Biaxial flexural strength(MPa)	Density(%)	Grain size(μm)
1780-5	1300-1	12.5±7	191±0.15	99.39	7.2
"	1300-5	13.6±9	163±0.18	99.63	8.4
"	1300-10	13.2±10	177±0.16	99.04	9.1
"	1300-20	11.9±8	130±0.11	99.78	10.1
1780-5	1600-1	13.4±8	137±0.17	98.54	8.5
"	1600-5	16.1±10	154±0.11	98.99	9.3
"	1600-10	14.4±11	150±0.14	99.43	10.0
"	1600-20	19.0±10	161±0.18	99.87	11.0

4. CONCLUSIONS

The results for application of the two-step sintering process to the transmittance alumina specimens fabricated with the powder injection molding method show that as the two-step sintering temperature and time increase, the grain size tends to slowly increase due to grain-boundary diffusion process. In particular, at 1600℃ and 20 hours, it was confirmed that the light transmittance increased a great deal due to the long-hour uniform grain growth without degradation of mechanical strength. At a low TSS temperature of 1300℃ and a low TSS holding time of 1 hour, however, the specimen showed the highest strength, 191 MPa, due to the grain size being below 5 μm, but the light transmittance was the lowest at 12.5%. As a result, transparent polycrystalline alumina with a high transparency and mechanical strength was obtained by elaborately controlling the two-step isothermal sintering steps.

ACKNOWLEDGEMENT

This research has been supported by the Energy Resource Technology Development Project, Ministry of Knowledge and Finance. The authors would like to express their thanks for the support.

REFERENCES

[1] Krell, A., Blank, P., Ma, H., Hutzler, T., van Bruggen, M. P. B. and Apetz, R., Transparent sintered corundum with high hardness and strength. J. Am. Ceram. Soc., 2003, 86, 12–18

[2] Wei, G. C. and Rhodes, W. H., Sintering of translucent alumina in a nitrogen–hydrogen gas atmosphere. J. Am. Ceram. Soc., 2000, 83, 1641–1648.

[3] O, Y. T., Koo, J. B., Hong, K. J., Park, J. S. and Shin, D. C., Effect of grain size on transmittance and mechanical strength of sintered alumina. Mater. Sci. Eng. A, 2004, 374, 191–195

[4] Apetz, R. and van Bruggen, M. P. B., Transparent alumina: a light-scattering model. J. Am. Ceram. Soc., 2003, **86**, 480–486.

[5] J.Y. Roh et al. , Novel fabrication of pressure-less sintering of translucent powder injection molded (PIM) alumina blocks, Ceram. International 37, 2011, 321–326.

[6] D.S. Kim, J. Lee, R.J. Sung, S.W. Kim, H.S. Kim, J.S. Park, Improvement of translucency in Al_2O_3 ceramics by two-step sintering technique, J. Eur. Ceram. Soc., 27, 3629–3632, 2007.

[7] C. Wang, Z, Zhao, Translucent $MgAl_2O_4$ ceramic produced by spark plasma sintering, Scripta Material 61, 193-196, 2009.

[8] S. R. Casolco et al., Transparent/translucent polycrystalline nanostructured yttria stabilized zirconia with varying colors, Scripta Material, 58, 516-519, 2008.

[9] W. J. Tropf, M.E. Thomas, in: E.D. Palik (Ed.), Handbook of Optical Constants of Solids II, academic press, 1991, p. 881

[10] Y. M. Chen, R. J. Smales, K. H. K. Yip, W. J. Sung, Translucency and biaxial flexural strength of four ceramic core materials, Dental Mater., 24, 1506-1511, 2008.

[11] C. J. Hwang, T. H. Kwon, A Full 3D Finite Element Analysis of the Powder Injection Molding Filling Process Including Slip Phenomena, Polym. Eng. Sci., 42, 33-50, 2002.

[12] I. W. Chen, X. H. Wang, Sintering dense nanocrystalline ceramics without final-stage grain growth, Nature 404, 168-171, 2000.

[13] Z. R. Hesabi, M. Haghighatzadeh, M. Mazaheri, D. Galusek, S.K. Sadrnezhaad, Suppression of grain growth in sub-micrometer alumina via two-step sintering method, J. Euro. Ceram. Soc., 29, 1371–1377, 2009.

Author Index

Author Index